The Gasoline Automobile

You are holding a reproduction of an original work that is in the public domain in the United States of America, and possibly other countries.You may freely copy and distribute this work as no entity (individual or corporate) has a copyright on the body of the work.This book may contain prior copyright references, and library stamps (as most of these works were scanned from library copies).These have been scanned and retained as part of the historical artifact.

This book may have occasional imperfections such as missing or blurred pages, poor pictures, errant marks, etc. that were either part of the original artifact, or were introduced by the scanning process. We believe this work is culturally important, and despite the imperfections, have elected to bring it back into print as part of our continuing commitment to the preservation of printed works worldwide. We appreciate your understanding of the imperfections in the preservation process, and hope you enjoy this valuable book.

Copyrighted by
P. M. HELDT
1918

ALL RIGHTS RESERVED

PREFACE

THE present volume is based on a series of articles which appeared in *The Horseless Age* during 1914 and 1915. The material was originally prepared with the intention that it should subsequently be published in book form, but various things interfered with the immediate carrying out of this plan. The author has now acquired the publishing rights to the series and has thoroughly revised it, adding new material especially to the section on Electric Lighting and Starting, and he offers it herewith as a volume of the set on The Gasoline Automobile.

Like the other two volumes of the set, the present volume was written mainly to meet the needs of automotive engineers and students of the science and art of automotive engineering. However, the third volume differs somewhat from the two preceding ones, both in method of treatment and in the scope of its appeal. There are in the country at present a large number of managers and employees of garages and service stations who have to take care of electrical apparatus, and it has been the author's aim to incorporate in the book such material as is required to make it serve them as a practical handbook of their trade. It was particularly with this object in view that a treatment of the elementary principles of electricity was included.

What will probably appeal most to the reader who is connected with a garage or service station are the sections on the maintenance, care and repair of the different items of electrical equipment. In dealing with these subjects the aim has been to make the information given of as wide application as possible, and always to explain the underlying reasons for defective operation observed and for remedies applied.

Most of the drawings with which this volume is illustrated were made by J. W. Kipperman, whose neat and accurate work will be appreciated by the reader.

The author will feel indebted to any of his readers who will apprise him of errors either in the text or the illustrations.

List of Chapters

I.—Principles of Electrical Phenomena	1
II.—Primary Electric Batteries	10
III.—Secondary or Storage Batteries	17
IV.—Magnetism	39
V.—Electromagnetic Induction	49
VI.—Measuring Instruments	62
VII.—The Magneto Generator	70
VIII.—Low Tension Ignition	76
IX.—Elements of a High Tension Battery System	83
X.—The Spark Plug	96
XI.—Vibrator Coil Ignition for Multi-Cylinder Motors	107
XII.—Modern Battery Systems	119
XIII.—Magneto and Coil Ignition	132
XIV.—The High Tension Magneto	143
XV.—Special Types of Magnetos—Care and Adjustment	153
XVI.—Methods of Spark Timing	166
XVII.—Combined Magneto and Battery Systems	177
XVIII.—Two Point Ignition	192
XIX.—Cables, Terminals, Switches and Wiring Methods	200
XX.—Storage Battery Charging—Magnet Recharging	211
XXI.—Ignition Testing Apparatus and Tests	226
XXII.—Electric Generators	235
XXIII.—Generator Control	259
XXIV.—Battery Switches—Charge Indicators	286
XXV.—Lamps and Fittings	296
XXVI.—Starting Motors	323
XXVII.—Starter Drives	336
XXVIII.—Fuses, Lighting Switches, Wiring Fixtures and Wiring	360
XXIX.—Battery Arrangement—Care and Maintenance of Systems	377
Appendix	391

CHAPTER I

Principles of Electrical Phenomena

Electricity is a form of energy—a medium for the transmission of power. Other forms of energy are heat, light, the chemical potential energy of certain materials such as coal, wood, petroleum, etc., and the kinetic energy of moving bodies. The most distinctive characteristics of electricity, as compared with other forms of energy, are that it can be readily and accurately controlled, and lends itself to transmission over long distances without undue loss. Various theories have been evolved to explain the exact nature of electrical phenomena. These, however, are of interest chiefly to the scientist, and do not directly concern the practical electrical worker. We will, therefore, not discuss them here. Most electrical phenomena can be clearly explained by considering electricity to be a fluid, and electric current to be a fluid in motion. In other words, electric phenomena may be explained by hydraulic analogies—a method which we shall use here.

Static and Current Electricity—A distinction has long been made between static electricity and dynamic electricity (or current electricity). As a matter of fact, there is only one kind of electricity, and the two terms may be regarded as applying to opposite extremes in the range of electrical phenomena commonly dealt with. We speak of electrostatics when referring to phenomena connected with frictional electric or induction machines. These machines produce exceedingly small quantities of electricity at extremely high pressures, and these two features are characteristic of so-called electrostatic phenomena. On the other hand, when we deal with any of the industrial sources of electricity we have comparatively large quantities of electric fluid at a relatively low pressure. However, in some automobile applications of electric currents (dynamic electricity) the pressure used is sufficiently high to produce what are ordinarily regarded as electrostatic effects, and these will be referred to again later on. For the most part, we have to do with the phenomena of electric currents.

Conductors and Insulators—All substances conduct electricity to some extent, but the conducting powers of different materials vary enormously. All of the metals are good conductors, the best among them being silver, copper, gold, aluminum, platinum, zinc, iron and nickel. Materials which have a very low conductivity—in other words, through which electricity passes with the greatest difficulty—are known as insulators. These insulators serve to keep the current in the conductors which they support or surround. However, there are no absolute insulators or materials which completely prevent leakage of current. The best insulators are glass, porcelain, mica, hard rubber, ebonite, silk, etc.

Difference of Potential—Water which is acted upon only by the force of gravity always flows from a higher to a lower level, if free to do so. We say that the flow from one point to the other is caused by the difference in level of these two points. Similarly, when an electric current flows in a conductor it flows from a point of high potential to a point of low potential, and the current flow is due to a difference in potential. Electric potential, therefore, corresponds to water level in hydraulic phenomena. Difference in electric potential is due to an electromotive force. The electromotive force in electrical phenomena corresponds to the force of gravity in hydraulic phenomena. As the force of gravity on any object (the weight of the object) is measured in pounds, so the electromotive force is measured in volts. In other words, the volt is the unit of electromotive force. The technical definition of the volt will be given later.

In Fig. 1, A and B represent two tanks containing water, which communicate through a pipe. Since the water stands at different levels in the two tanks, water will flow from the higher to the lower tank through the pipe. The difference in level is designated by a. In Fig. 2 are shown the same two tanks connected by a pipe of the same length and diameter of bore as in Fig. 1, but in this case the difference in level b is twice as great as a in Fig. 1. The result will be that water will flow from the higher to the lower tank twice as fast as in Fig. 1. This is expressed technically by saying that the rate of flow is directly proportional to the difference in level.

In electrodynamic phenomena the current corresponds to the rate of flow in hydraulic phenomena, and in any electric circuit, if the electromotive force is varied, other things remaining the same, the current will vary in the same proportion as the electromotive force. The unit of current is the ampere, a definition of which will be given later on.

In the hydraulic analogy cited, it was stated that in the

PRINCIPLES OF ELECTRICAL PHENOMENA. 3

two cases the length and diameter of bore of the connecting pipe remained the same. If the dimensions of the pipe were varied while the difference in level remained the same, then the rate of flow would also vary, because with a change in the dimensions of the pipe there would be a change in the frictional resistance encountered by the flowing water. If the pipe be lengthened, the resistance naturally would be increased, whereas if the bore of the pipe be increased, the resistance would be lessened. The electrical conductors composing the circuit through which the electric current flows also offer a certain resistance to its flow. This resistance varies directly as the length of the conductor, inversely as the cross-sectional area of the conductor and inversely as the "conductivity" of the material of which the conductor is composed. That is to say, if we double the length of the conductor we

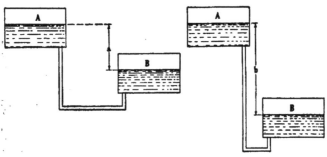

FIG. 1.   FIG. 2.

double its resistance, but if we double its cross section or use a material of twice the conductivity we halve the resistance.

Ohm's Law—These three important factors in electrical work, the electromotive force, the current and the resistance, are connected by a very simple law, named after its discoverer, Ohm's law, which is to the effect that the current is equal to the electromotive force divided by the resistance. Mathematically this is expressed as follows:

$$\text{Current} = \frac{\text{Electromotive force}}{\text{Resistance}}$$

or in abbreviated form,

$$C = \frac{E}{R}$$

By transforming this equation we find that

Electromotive force = current $\times$ resistance

4   PRINCIPLES OF ELECTRICAL PHENOMENA.

and

$$\text{Resistance} = \frac{\text{Electromotive force}}{\text{Current}}$$

Therefore, by means of this law, if we know any two of the factors, we can readily find the third. As already stated, the electromotive force and the current are measured in volts and amperes respectively, and to this it may now be added that the resistance is measured in ohms. One ohm is the resistance of a column of mercury one square millimeter in cross section and 106.3 centimeters in length at the temperature of melting ice (32 degrees Fahrenheit). It is necessary to specify the temperature because the resistance of mercury, as well as of most other conductors, increases with the temperature.

Chemical Effects of Current—The definition of the ampere, the unit of current, is based on the chemical effects of the current. When two wires are connected to a source of electric current and their free ends are dipped into a glass of slightly acidulated water, the current will pass through the water from one wire to the other, and gas bubbles will be seen rising from the surface of the water where the wires enter. The gases given off are hydrogen and oxygen, the constituents of water, which later is decomposed by the passage of the current. The oxygen rises at the wire through which the current enters (or, rather, is supposed to enter) the water, and the hydrogen at the wire through which the current leaves (or is supposed to leave) the water. The former wire, of course, connects to that terminal or pole of the source of current from which the current proceeds, which is known as the positive pole, and the latter to the pole at which the current returns, which is known as the negative pole. As water is composed of two parts (by volume) of hydrogen to one part of oxygen, gas will be given off much faster at the wire by which the current leaves the water than at the other, and a device of this kind, therefore, can be used for determining the direction of the current. The portions of the wires extending into the water are known as electrodes. The reason for adding a small amount of acid to the water in the above experiment is that it increases the conductivity of the water, pure water being a very poor conductor.

Electrolytes—Conductors which in conducting a current are decomposed by it are known as electrolytes. Most electrolytes are solutions of metallic salts. One of the most familiar electrolytes is a solution of copper sulphate (blue vitriol). If current is sent through such a solution metallic copper is deposited on the electrode by which the current leaves and

PRINCIPLES OF ELECTRICAL PHENOMENA.

gases accumulate at the other electrode. The rate at which copper is deposited depends upon the current flow, and this permits of defining the unit of current, the ampere. One ampere is a current which during one second deposits 0.00113 gram of silver from a bath of silver nitrate in water. This is the standard definition of the ampere, which, like all other electrical units, is based upon the metric system. A deposit of 0.00113 gram per second is equal to 0.009 lb. per hour. The figure 0.00113 is termed the electro-chemical equivalent of silver. Each metal has its own electro-chemical equivalent, but some, like copper and iron, which form two different classes of compounds, have two different electro-chemical equivalents each.

Definition of the Volt—Having in the above given definitions of the unit of current, the ampere, and of the unit of resistance, the ohm, Ohm's law now enables us to give the following definition of the unit of electromotive force, the volt: A volt is an electromotive force which when applied to the ends of a conductor having a resistance of one ohm will cause a current of one ampere to flow through that conductor.

Specific Resistance—Every conducting material has what is known as a specific resistance. In the metric system this is the resistance in ohms between opposite faces of a cube of the material whose sides are 1 centimeter long. Most conductors used in practical work are in the shape of circular wires and in order to be able to quickly calculate the resistance of a wire from its dimensions and the constant corresponding to the material of which it is composed, it has become customary to express the cross section of a wire in circular mils. A round wire 0.001 inch in diameter has a cross section of one circular mil, and to find the cross section of any other wire in circular mils, we need only square the diameter of the wire in thousandths of an inch. For practical purposes, therefore, instead of taking the resistance between opposing faces of a one centimeter cube as the specific resistance, it is better to use the resistance of a circular wire 0.001 inch in diameter and 1 foot long. Such wires of the more common metals have the following resistances at 32 degrees Fahr.:

	Ohms
Silver	9.048
Copper	9.612
Gold	12.38
Aluminum	17.53
Iron	58.45
Nickel	93.86

The figures here given are for pure, annealed metal and the resistance of commercial metals would be somewhat higher. These figures permit of rapidly calculating the resistance of any wire of known dimensions. For instance, let it be required to find the resistance of 1,000 feet of copper wire 0.032 inch in diameter. The resistance is evidently

$$\frac{9.612 \times 1000}{32 \times 32} = 9.39 \text{ ohms}.$$

Temperature Coefficient—The resistances thus found are those at 32 degrees Fahr. At higher temperatures the resistances of practically all metals are greater, increasing by a certain fixed proportion for each degree of temperature increase. The proportional increase, of course, is different for each metal. The temperature coefficients or coefficients of increase in resistance per degree Fahrenheit for the different metals are as follows:

Copper	0.0024
Silver	0.0022
Gold	0.0021
Aluminum	0.0024
Iron	0.0030
Nickel	0.0035

These figures in conjunction with those given above enable us to determine the resistance of any wire of given dimensions at any temperature. For instance, if we take the copper wire of the previous example which at 32 degrees Fahrenheit had a resistance of 9.39 ohms, at 82 degrees Fahr., which is 50 degrees higher, this same wire will have a resistance of

$$9.39 + (9.39 \times 50 \times 0.0024) = 10.52 \text{ ohms}.$$

A rough rule for the increase in the resistance of copper wire is one per cent. for every 4 degrees Fahr. increase in temperature.

In this country copper wire is always made according to the Brown & Sharpe gauge.

Resistance Calculations—Wherever an electric current flows it encounters resistance. For instance, if we connect the poles of a battery cell by a length of wire, the current which will flow through the wire will be determined not solely by the resistance of the wire but by the internal resistance of the cell as well. Thus in Fig. 3 suppose the cell has an electromotive force of 2 volts, that its internal resistance is 0.1 ohm and that the resistance of the wire connecting its terminals is 1 ohm. Then the total resistance in circuit evidently is

$$1 + 0.1 = 1.1 \text{ ohms}$$

and the current that will flow in accordance with Ohm's law, is

PRINCIPLES OF ELECTRICAL PHENOMENA. 7

$$\frac{2}{1.1} = 1.82 \text{ amperes.}$$

Conductors through which the same current flows successively are said to be in series and their resistances may be added together.

Now, consider that the terminals of the cell are connected by two wires, as shown in Fig. 4, one having a resistance of, say, 2 ohms and the other of 3 ohms. Such wires are said to be connected in parallel. To find the effective resistance of these two wires in parallel we proceed as follows:

Suppose for a moment that an electromotive force of 1 volt is effective between the terminals A and B of the cell when the circuit is completed and the current flows. It is obvious that the current in one wire would be ½ ampere and

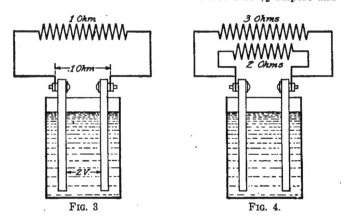

FIG. 3.   FIG. 4.

in the other ⅓ ampere. The total current therefore would be

$$½ + ⅓ = ⅚ \text{ amperes}$$

and since

$$\text{Resistance} = \frac{\text{Electromotive force}}{\text{Current}}$$

and the electromotive force is 1 volt, the equivalent resistance of the two wires is evidently

$$\frac{1}{⅚} = \frac{6}{5} = 1.2 \text{ ohms.}$$

and the total resistance in circuit

$$1.2 + 0.1 = 1.3 \text{ ohms.}$$

Force, Power and Energy—In studying electrical phenomena it is very important that we have a clear conception of

8   PRINCIPLES OF ELECTRICAL PHENOMENA.

the terms force, power and work (or energy). Force may be defined as the ability to do mechanical work. A body resting on a table exerts a force of so many pounds (its weight) on the table. If we wish to raise the body to a higher position, then we must apply to it a force equal to that with which it presses against the table and the product of the force in pounds into the distance it is lifted in feet is the work done in foot-pounds. For instance if a 10-pound weight is lifted through a vertical distance of 10 feet then the work done is

$$10 \times 10 = 100 \text{ foot-pounds.}$$

Power is the rate of doing work. If the 10-pound weight referred to were raised at the rate of 5 feet per minute then the power would be equal to

$$10 \times 5 = 50 \text{ foot-pounds per minute.}$$

Electric Energy—In explaining the factors that make up electric power, it is again well to refer to an hydraulic analogy. The power of a waterfall, for instance, depends upon two factors, viz.: the amount of water passing per minute and the height of the fall, which latter determines the pressure with which the water will impinge upon the blades of a water wheel, say. Similarly, in the case of a steam boiler, the power depends upon the amount of steam given off in a unit of time and upon the pressure of the steam. We may therefore say that the two factors of power are pressure and quantity per unit time (or rate of flow). The same holds true in electrical work. The pressure in this case is the electromotive force, measured in volts, and the rate of flow is the current measured in amperes. When a current of one ampere is flowing under a pressure of one volt, we have a unit of electrical power known as one watt. In general,

$$\text{Electric Power} = \text{Electromotive force} \times \text{Current}$$

or

$$\text{Watts} = \text{Volts} \times \text{Amperes.}$$

In this connection, it may be pointed out that 746 watts is equal to one horsepower or 33,000 foot-pounds per minute. Electrical engineers use, instead of the horsepower, the kilowatt, which is equal to 1,000 watts, or about $1\frac{1}{3}$ horsepower.

If we multiply the power or rate of doing work by the length of time the power is applied, we get the amount of work done. The unit of electrical work, known as the joule, is equal to one watt for one second. This unit, however, is rarely used in practical work. One foot-pound (unit of mechanical work) equals 1.356 joules (electrical unit of work).

Heat Losses—It is a well-established law of nature that energy is indestructible. When electrical energy is produced by any of the usual means, energy of some other form, the exact equivalent of the electrical energy produced, disappears. Again, the electrical energy exists only while the current flows and is converted immediately into some other form of energy. When we connect a wire across the terminals of an electric battery, the wire will increase in temperature, showing that heat is being generated. In fact, all of the energy represented by the electric current in that case is converted into heat. We found that the power of the electric current was equal to the product of the electromotive force, into the current. But inasmuch as the electromotive force, according to one version of Ohm's law, is equal to the product of the current into the resistance which it overcomes, the rate of energy or heat production is equal to the current in amperes multiplied by itself and by the resistance in ohms. This may be written,

$$L = C^2R \text{ watts.}$$

In all applications of electric energy, except for heating purposes or where the heat is indirectly utilized; as in incandescent lamps, etc., this heat represents a loss, and it is therefore common among electrical men to speak of the C^2R losses which means the resistance or heat losses in the conductors.

CHAPTER II

Primary Electric Batteries

The oldest source of current electricity is the primary electric cell. In its simplest form an electric cell consists of a vessel containing a solution of sulphuric acid in which are immersed a rod of zinc and a rod of copper or gas carbon. The zinc is known as the negative and the copper or carbon as the positive element of the cell. When the terminals of the positive and negative elements are connected by a wire outside the cell, as shown in Fig. 5, a current will flow from the positive element (copper) to the negative element (zinc) outside the cell, and from the negative element (zinc) to the positive element (copper) inside the cell. As the current flows, the zinc rod is eaten away, the zinc combining with the sulphate radical of the sulphuric acid to form zinc sulphate, which is held in solution, and the remaining element of the sulphuric acid, hydrogen, accumulating on the copper or carbon element and possibly bubbling up through the solution. There are many other metals besides zinc and copper which can be used for the elements of primary cells, but these two have proven the most practical. The element which is consumed the most rapidly is always the negative element.

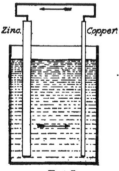

FIG. 5.

Rule for Direction of Current—A good rule to remember for direction of current flow is that in a cell or any electrolytic bath the current always flows in the direction in which the metal travels. In the above described simple primary cell the metal passes from the zinc element into the solution and the current passes from the zinc element through the solution to the positive element.

Consumption of Zinc—According to the law of the conservation of energy, the production of electric energy in the cell must be accompanied by the loss of energy in some other form. In this case it is derived from the chemical potential energy of the zinc.

FIG. 6 — CONVENTIONAL DIAGRAM OF A BATTERY.

When the zinc from the negative element of the cell combines with the sulphate radical of the sulphuric acid it practically burns and gives out energy in the same way as coal does when burned, except that in this case the energy appears in the form of electricity instead of as heat. A part of the heat of combination of the zinc sulphate is used up in splitting up the sulphuric acid into hydrogen and "sulphate," and the rest appears as electric energy in the circuit. One pound of zinc will furnish 375 ampere-hours, that is, say, one ampere for 375 hours.

Polarization—The hydrogen component of the sulphuric acid accumulates in small bubbles on the copper or carbon (positive) element of the cell. It thereby increases the internal resistance of the cell and also sets up a counter-electromotive force, because the hydrogen is negative with respect to the zinc. The result of this effect, which is known as polarization, is that after the cell has been delivering current for a short time the current falls off in value. Owing to this fact the simple primary cell can be used only for delivering current intermittently for short periods as required in operating house bells, for instance.

A number of cells electrically connected together are known as a battery. In electrical discussions it is usual to represent a battery cell diagrammatically by means of one short and heavy and one long and thin line, the two lines being parallel (Fig. 6). The heavy lines represent the zinc or negative elements and the light lines the carbon or positive elements. The positive pole or terminal of a cell or battery is generally indicated by a plus (+) sign, and the negative terminal by a minus (—) sign.

Methods of Connecting Cells—There are three methods of electrically connecting up battery cells. If in a group of cells the positive pole of one cell is always connected to the negative of another cell, as shown in Fig. 7, the cells are said to be connected in series. In a series connected battery the electromotive force between the end terminals is equal to the product of the number of cells into the electromotive force of a single cell. The maximum current obtainable from such a battery, however, does not exceed that obtain-

12 PRIMARY ELECTRIC BATTERIES.

able from a single cell. Another way of connecting up a set of cells is shown in Fig. 8, and is known as parallel or multiple connection. In this case all of the positive terminals are connected together and all the negative terminals likewise. The effective electromotive force of a multiple battery is the same as that of a single cell, but, naturally, a much larger current can be drawn from such a battery than from a single cell, as each cell delivers only a fraction of the total current. Of course, only cells of the same electromotive force should be connected in parallel, as if one cell had a higher electromotive force than the others with which it was connected in parallel it would send a current through the latter in a reverse way. With the electromotive forces of all cells exactly alike they balance each other and there can be no reverse current through any of them.

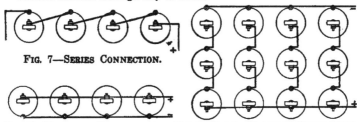

FIG. 7—SERIES CONNECTION.

FIG. 8—PARALLEL CONNECTION.   FIG. 9—SERIES-MULTIPLE CONNECTION.

Electromotive Force Independent of Size—It may here be pointed out that the electromotive force of a battery cell depends only upon the materials of the elements and the solution and not in any way upon the size of any part of the battery or of the battery as a whole. However, the internal resistance of a cell, and consequently the maximum current which can be drawn from it, depend upon the form and size of the parts of the cell. The resistance is the less the larger the surface area of the elements and the closer they are together.

Series-Multiple Connection—There is a third way of connecting up cells, known as the series-multiple connection, which is a combination of the two methods previously described. It is illustrated in Fig. 9. If we have twelve cells, for instance, and want to use them for a purpose that calls for an electromotive force equal to four times that of a single cell, we can connect the cells in three groups of four in series. Each cell will then be delivering only one-third the total current required. In a series-multiple battery the electromotive force is equal to that of a single cell multiplied by

PRIMARY ELECTRIC BATTERIES. 13

the number of cells connected in series and the maximum current capacity is equal to that of a single cell multiplied by the number of parallel groups.

Depolarizers—In order to prevent polarization of cells and thus enable them to give a steady current continuously, many forms of depolarizers have been employed in primary cells. The object of the depolarizer is to furnish an element or radical combining with the hydrogen or its equivalent as fast as it is formed, so preventing its adherence to the positive plate, which latter constitutes the phenomenon known as polarization.

Wet Cells—Primary cells may be divided into wet cells and dry cells. The former, as a rule, consist of a jar of insulating material, often glass, with a cover of insulating material, usually a molded composition, which holds the two elements of the cell. The jar is filled with solution to near the top. Sometimes the positive element is enclosed in a porous cup at the center of the jar and two different solutions are used, one in the porous cup and one in the jar, the former acting as depolarizer. Wet cells are never used on automobiles on account of the danger of the acid solution slopping over, and we, therefore, need not discuss them in detail. The only type of primary cell used on automobiles is the so-called dry cell which is constructed as follows (Fig. 10):

Dry Cells—The negative element of a dry cell consists of a zinc cup A which also forms the jar or container for the whole cell. In the center of this zinc cup is located a carbon rod B, which is supported by means of an insulating composition C, forming a closure for the zinc cup. The carbon rod does not extend entirely to the bottom of the zinc cup. The electrolyte or exciting fluid consists of a solution of sal-ammoniac (chloride of ammonia), which is held by some absorbent material D, like blotting paper or strawboard, with which the interior of the zinc cup is lined. In between this absorbent lining and the central carbon rod is a space which is filled with a depolarizing agent E. The latter generally consists of black oxide of manganese. It is mixed with powdered or granulated gas carbon and the whole is saturated with exciting fluid.

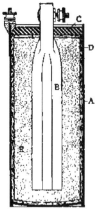

Fig. 10—Dry Cell in Section.

A considerable amount of electrolyte is required because the sal-ammoniac is consumed at a somewhat faster rate than the zinc, one pound of sal-ammoniac giving 225 ampere-hours.

Black oxide of manganese, the depolarizer, readily gives off some of its oxygen, which combines with the negative element of the electrolyte after it has given up its chlorine to the zinc in forming zinc chloride. Owing to the provision of this depolarizer a fairly large current can be drawn from a dry cell continuously.

Standard dry cells as used for ignition and similar purposes are 6 inches high and 2½ inches in diameter. To increase the surface area of the carbon rod it is sometimes made flat or corrugated instead of cylindrical. The electromotive force of each cell is about 1.5 volts, and when a cell is short circuited—that is, when its two terminals are connected by a short, heavy wire, or through an ampere meter, which is equivalent thereto—it will furnish a current of 15 to 22 amperes, according to its condition. As in that case the current depends only upon the internal resistance of the cell it follows that the internal resistance varies from

$$\frac{1.5}{22} = 0.08 \text{ ohm to } \frac{1.5}{15} = 0.1 \text{ ohm.}$$

Of course, short-circuiting is injurious to the cell and the short circuit should be maintained only for an instant to determine the cell's condition. Cells should not be tested unnecessarily—only when there is occasion for it.

Causes of Deterioration—The top of the cell, as already stated, is covered by an insulating composition which is poured in place while in the molten state. This covering is depended upon not only to insulate the two elements from each other, but also to prevent evaporation of the electrolyte. It is this evaporation which usually causes the deterioration of dry cells, as with a loss of electrolyte the internal resistance of the cell increases. Cells, therefore, deteriorate slowly even while not in use, and it is always a good plan when purchasing dry cells to make sure that they are fresh and in good condition. Dealers now make it a rule to apply an ammeter to every cell as they are handing them out, demonstrating to the customer that they are fresh. Dry cells should always be stored in a dry, cool place, so that evaporation will be a minimum and there is no chance of moisture settling on the insulating covering and causing leakage paths which will gradually drain the cell of its charge. When a battery of dry cells is used on a car they should be carefully packed in a wooden box or a box lined with insulating material so they

PRIMARY ELECTRIC BATTERIES.   15

cannot shake about. In no case should the battery be placed directly in a steel box, as, in spite of the cardboard covering, the binding posts on the zincs would come in contact with the walls of the box and short-circuit the cells, quickly ruining them. In installing dry cells on a car, care should be taken to place them where they cannot possibly be drenched while the hose is turned on the car in washing.

Efficiency of Series-Multiple Connection—When dry cells are used for a purpose where considerable current is required, as in lighting, it is always well to use a so-called multiple battery. The number of cells to be connected in series, of course, depends upon the voltage required. For instance, if the lamps are 6 volt, then five cells in series are required, for although these have a voltage of 7.5, a loss of 1.5 volts in the cells themselves and in the wiring can be figured on. In order to insure high efficiency and long life of the cells, several such series of five should be connected in multiple. How this affects the efficiency may be shown by an example.

Suppose two 2-candle power 6-volt lamps are to be lighted by means of dry cells. These together would take a current of about 1 ampere. Let us first suppose that a single series of five cells is used, each of an electromotive force of 1.5 volts and an internal resistance of 0.01 ohm. Then the total electromotive force is 7.5 volts and the total resistance 0.7 ohm. The battery will deliver

$$7.5 \times 1 = 7.5 \text{ watts},$$

whereas the loss on account of the internal resistance of the battery will be

$$1 \times 1 \times 0.5 = 0.5 \text{ watt}.$$

Hence the efficiency will be

$$\frac{(7.5 - 0.5)\ 100}{7.5} = 93 \text{ per cent.}$$

On the other hand, if we use four series of five cells each in parallel, the internal resistance of the whole battery will only be one-fourth as much or 0.125 ohm, and the loss in the battery will be

$$1 \times 1 \times 0.125 = 0.125 \text{ watt}.$$

Hence the efficiency in this case will be

$$\frac{(7.5 - 0.125)\ 100}{7.5} = 98 \text{ per cent.}$$

These figures, however, hardly reflect the full advantage gained by using a multiple battery.

Wireless Battery Holders—The two elements of a dry cell are provided with a binding screw each, and usually connections are made by means of short connectors made of

16 PRIMARY ELECTRIC BATTERIES.

stranded insulated copper wire with a stamped copper terminal at each end. However, the nuts of the binding screws have a tendency to rattle loose on a car, and to prevent the trouble resulting therefrom and also to facilitate the renewal of cells, so-called wireless battery holders are now sometimes used. A sketch of the Patterson wireless battery holder is shown in Fig. 11. It consists of a number of units in vulcanite, each supplied with a socket fitted with a threaded cap and "bridge." This bridge consists of a flat strip which normally rests against a U-shaped piece and is out of contact with a circular brass pad below it. Bridges and caps are connected in series. Special cells are used with this holder, which have the top end of their zincs threaded to fit the threaded cap on the holder, and the usual project-

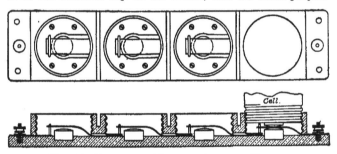

FIG. 11—PATTERSON WIRELESS BATTERY HOLDER.

ing carbon blocks with binding screws are replaced by a circular brass pad terminal, which, when the cell is screwed in place on the holder, lifts the bridge and makes metallic contact between the pad on the cell and the pad in the holder. Thus connections are automatically made by screwing the cells into the holder. If one cell should be removed for any reason, the bridge will move up against the U-shaped piece, thus keeping the circuit closed.

Reviving Old Dry Cells—When dry cells have become exhausted from "drying up," they may be revived by boring two or three holes (about $3/8$ inch) through the insulating covering and into the depolarizing material and filling the holes with water or a solution of sal-ammoniac, allowing the latter to soak in and renewing it a number of times. Then seal the holes up either with the borings from the insulating material or with paraffin wax in the molten state. However, dry cells are now sold at such a low price that no material saving can be expected from thus reviving old cells.

CHAPTER III

Secondary or Storage Batteries

If two plates of sheet lead are placed in a bath of sulphuric acid solution and connected to a source of electric current, and if, after the current has been flowing for a while, the plates are disconnected from the electric source and connected instead to an electric indicator, it will be found that the cell of which they form the elements has itself become a source of electric current. Current will pass through the cell in a direction opposite to that in which it flowed when the cell was connected to the outside source of current. Such a cell forms a simple secondary or storage cell which can be "charged" by sending a current through it and which will "discharge" or deliver a current when its elements are connected by a wire, etc. The effect of sending a current through the cell is to oxidize the surface of one of the lead plates. Lead peroxide is formed by the action of the current on the plate by which the charging current enters the cell (the positive plate), while the surface of the other (the negative) plate is at the same time converted into spongy lead. To obtain a storage cell of such capacity as to be of any practical use it is necessary to so form the elements that they present an enormous surface to the action of the electrolyte in proportion to their bulk or weight, the charging and discharging operations must be repeated a number of times (this process being called forming), and the plates may be subjected to a preliminary forming process consisting in dipping them into nitric acid.

Storage cells made up of elements of metallic lead, which are formed electrically as above described, are known as Plante or electrically formed cells. They have a long life but are practically never used on automobiles on account of their great weight for a given capacity.

Chemical Actions—Modern storage cells as used on gasoline vehicles are of the Faure or pasted plate type. In these the electrodes or plates consist of grids of lead (usually antimonious lead, an alloy exceeding pure lead in hardness), to which pastes made up of lead salts, sulphuric acid solution, and certain ingredients increasing their adhesive qualities, are applied mechanically. The paste for the positive plates is made of red lead or minium (Pb_3O_4) and the paste for the negative plates of litharge or lead monoxide (Pb O). The charging current converts the red lead on the positive plates into lead peroxide (Pb O) and the litharge on the negative plates to spongy metallic lead. During the discharge the spongy lead of the negative plates and the lead peroxide of the positive plates are both converted into lead sulphate. The chemical changes during both charge and discharge may be represented as follows:

DISCHARGE⟶

$$Pb\ O_2 + 2\ H_2 S\ O_4 + Pb = 2\ H_2 O + 2\ Pb\ S\ O_4$$
Lead Peroxide + Sulphuric Acid + Lead = Water + Lead Sulphate

⟵CHARGE

This chemical equation shows that the sulphuric acid (H_2SO_4) of the electrolyte takes part in the chemical reaction. During discharge some of the sulphuric acid of the electrolyte is split up and combines with the active materials (lead peroxide and spongy lead) to form water and lead sulphate. Consequently, there is less sulphuric acid in the electrolyte when the cell is discharged than when it is charged.

When the cell is in the charged state the active material on the positive plate (lead peroxide) has a chocolate brown color and the active material on the negative plate (spongy lead) a gray color. During discharge the active material on the positive plate assumes a lighter brown color and that on the negative plate becomes slate gray. In other words, both sets of plates become lighter in color during discharge and darker during charge.

Grids—One of the chief differences between storage batteries of the different makes is in the form of their grids. These serve the purpose of a support for the active material and of a current conductor. In designing them the aim is to make them as light and strong as possible, and so as to hold the active material securely and to enable them to withstand its expansion and contraction during charge and discharge. Two typical grids are shown in Fig. 12. Both consist of double lattice work between which the active material is incorporated. In one, known as the Diamond grid, the lattice work on opposite sides is so arranged that the points of the diamond do not come opposite, and the active material is locked between the crosses. In the other grid, the members run horizontally and vertically, and those on opposite sides are also staggered. In one make of storage battery the positive plates consist of perforated tubes of hard rubber filled with active material and a lead core joined to lead bars at top and bottom.

The thickness of the plates varies from $\frac{1}{8}$ to $\frac{3}{8}$ inch, according to the use to which the battery is to be put. Where the rate of discharge is low (ignition battery) the plates are made thick, whereas when the rate of discharge is high (starting battery) the thinnest plates are used. To permit high rates of discharge, a specially porous active material is sometimes used.

Positive and Negative Groups—Individual plates are as-

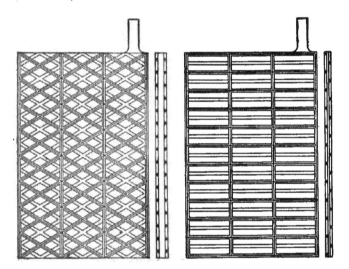

FIG. 12—TWO DESIGNS OF GRIDS.

sembled into groups by means of lead connecting straps which are burned (process of lead burning) to lugs formed on the grids. A group of positive plates and a group of negative plates are then assembled into an element so that each positive plate is located between two negative plates, the negative group always containing one more plate than the positive group. The reason for this is that the actions of expansion and contraction during charge and discharge are greatest in the positive plates and unless these actions are equal on both sides of the plate there is apt to be serious buckling or

20   SECONDARY OR STORAGE BATTERIES.

warping. With one more negative than positive plates, all of the latter are acted on equally on both sides and the danger of buckling is minimized.

Construction of Cell—The elements are placed in a cell or jar which is generally made of hard rubber. A typical vehicle lighting battery is the Willard of which a sectional view is shown in Fig. 13. Referring to the cut, the jar A is pro-

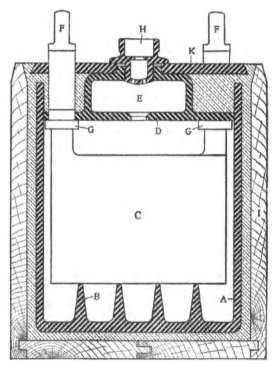

FIG. 13—WILLARD STORAGE CELL IN SECTION.

vided with bridges B on the inside at the bottom, on which the element C rests and which provides a considerable "mud space" at the bottom of the cell, where sediment shed by the plates can accumulate before it begins to short-circuit the cell by coming in contact with plates of opposite sign. The plates themselves must be placed rather close together so as to reduce the internal resistance of the cell to a minimum,

SECONDARY OR STORAGE BATTERIES.   21

yet there must be no possibility of their coming into contact with each other. To this end separators of some insulating material are usually placed between adjacent plates. These consist of perforated hard rubber or wood, the latter receiving a special treatment to make them acid-proof. These separators must not interfere with the free circulation of the electrolyte, as otherwise the electrical action would be hampered. Often both rubber and wood separators are used together, the rubber separators being perforated and the wood separators having one side grooved, their flat side being placed against the negative plates.

The cell is filled with an electrolyte consisting of a sulphuric acid solution made up of three or four parts by volume of distilled water to one part of pure sulphuric acid (different manufacturers using slightly different proportions) to ½ inch above the tops of the plates. The top of the cell is then closed by means of a hard rubber plate, the joints being sealed with a compound which softens when slightly heated. In the case of the Willard cell (Fig. 13), the covering D of the cell is formed with an expansion chamber E to take care of changes in volume of the solution during charge and discharge. The terminals F of the cell, which are formed integral with the connecting strips G, extend through cover D. They are covered with a layer of Para rubber, vulcanized directly to the corrugated surface of the terminal to prevent creeping of acid. A vent plug H of hard rubber, through which gases can escape, is screwed into the cover plate.

The complete cells are placed in a hardwood box I with dove-tailed joints, a hard rubber plate K is placed on top, held in place by the vent plug H, and the space between the cell and the wood box is filled with plastic material.

Another type of portable storage cell, the Exide, is shown in sectional view in Fig. 14. It has a composition cell cover with a downwardly flanged edge which when the cover is put in place on the cell forms with the top edge thereof a groove which is filled with sealing compound. A liquid tight joint between the terminals and the cell cover is made by means of a gasket and sealing nut. The cell sets directly in the wooden box.

The separators usually extend some distance above the plates, to prevent the possibility of short-circuits by conducting material bridging the plates on top, and they are prevented from floating up and thus uncovering the lower sections of the plates, by means of "hold down" bars fastened to the under side of the connector bars.

Voltage of Cell—The average electromotive force during

charge is 2.3 volts and the average during discharge is 2 volts. The electromotive force increases continuously during the charge and decreases during discharge. It attains a maximum value during charge of about 2.55 volts, but 10 minutes after charging ceases, it has dropped to about 2.1 volts. It is at this voltage that the discharge begins and the latter should never be continued after the voltage of the cell has dropped to 1.7, if the cell is not to be injured. The varia-

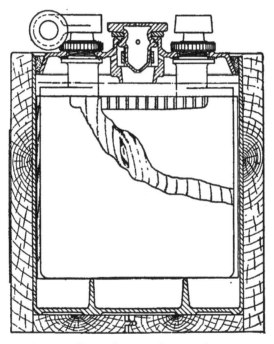

FIG. 14—EXIDE STORAGE CELL IN SECTION.

tions in voltage throughout the charge and discharge periods are represented by the curves in Fig. 15. The time in hours is plotted along the horizontal axis, and the voltage of the cell along the vertical axis.

Capacity Ratings—Storage batteries for ignition, lighting and starting are generally put up in wooden boxes with straps or handles to make them easily portable. The voltage employed in these services is frequently 6 and the cells are put

SECONDARY OR STORAGE BATTERIES. 23

up in batteries of three connected in series, though for lighting and starting, 12-volt cells are also much used. The usual capacities are 80, 100 and 120 ampere-hours, these being the rated discharge capacities. The following rules for rating lighting and combined lighting and starting lead batteries have been issued by the Society of Automotive Engineers:

"Lighting batteries shall be rated at the capacity in ampere-hours of the battery when discharged continuously at a 5-ampere rate to a final voltage of 1.8 per cell, the temperature of the battery beginning such discharge being 80 degrees F.

"Batteries for combined lighting and starting service shall have two ratings, of which the first shall indicate the light-

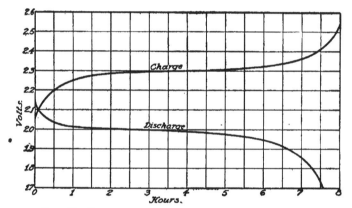

FIG. 15—CHARGE AND DISCHARGE CURVES OF LEAD CELL.

ing ability and be the capacity in ampere-hours of the battery when discharged continuously at a 5-ampere rate to a final voltage of 1.8 per cell, the temperature of the battery beginning such discharge being 80 degrees F. The second rating shall indicate starting ability and shall be the rate in amperes at which the battery will discharge for 20 minutes continuously to a final voltage of not less than 1.65 per cell. The temperature of the battery beginning such discharge to be 80 degrees F."

Standard Dimensions—The following standards for the dimensions of batteries for lighting and for combined lighting and starting have been issued by the Society of Automotive Engineers:

"The overall width of the battery, measured from side to side of case, shall not exceed 7½ inches.

"The overall height of the battery measured from bottom of case to top of handles shall not exceed 9½ inches.

"The overall length of the battery, measured from end to end of case, including handles, shall vary according to the capacity of the battery and its details of design. Handles shall, as standard, be placed at the ends of the battery, and provision for hold-down devices shall, as standard, be made at the ends of the battery. The space occupied by such handles and hold-down devices shall be in the direction of the length of the battery only, and not in the direction of its width. Terminals and connections shall not extend above the handles, the latter to be the higher point."

A 6-volt 100 ampere-hour lighting battery weighs about 60 pounds, and other 6-volt batteries in proportion. A 12-volt battery of the same watt-hour capacity as a 6-volt battery, weighs about 20 per cent more.

Variation of Capacity with Discharge Rate—For a given battery the actual capacity, of course, varies with the rate of dicharge, the higher the rate the lower the capacity. Fig. 16

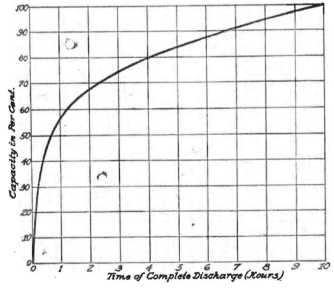

FIG. 16—VARIATION OF CAPACITY WITH RATE OF DISCHARGE.

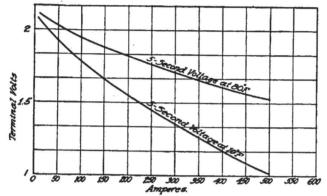

FIG. 17—VARIATION OF TERMINAL VOLTAGE WITH DISCHARGE RATE AT MODERATE AND LOW TEMPERATURE.

graphically shows the approximate relation between rate of discharge and capacity, the capacity at a 10-hour rate being taken as 100%. The capacity also varies with the temperature of the cells. If we call the capacity at 70 degrees Fahr. 100%, then the capacity at 20 degrees is only 62%, the decrease in capacity between these limits being in direct proportion to the drop in temperature. Above normal atmospheric temperature, while the capacity continues to increase, its rate of increase declines and the maximum possible increase above that at 70 degrees is only about 10 per cent.

Variation of Voltage with Temperature and Discharge Rate.—The terminal voltage of a cell is also greatly dependent on the temperature and this relation has been investigated by O. W. A. Oetting, who reported his results in a paper read before the Cleveland section of the S. A. E. on February 15, 1918. Fig. 17 shows the "five-second voltage" curves of a 100 amp.-hr. battery at 80 degrees Fahr., and at 10 degrees Fahr. By "five-second voltage" is meant the voltage that a battery will give at the end of five seconds at a certain discharge rate of current. In other words, the cell whose voltage is shown in Fig. 17 will have a voltage of 1.61 after five seconds' discharge at 400 amp. at a temperature of 80 degrees Fahr.

Determining State of Charge—The majority of automobile storage batteries are now charged automatically by means of an electric generator on the car. It is essential, however, that the battery be always kept in a fair state of charge, as otherwise it will deteriorate. There are two methods of determining

26 SECONDARY OR STORAGE BATTERIES.

FIG. 18—TYPICAL IGNITION BATTERY.

the state of charge of a cell, viz.: by means of a voltmeter and by means of an hydrometer. From Fig. 15 it was seen that the voltage of the cell gradually drops during the discharge, so that a voltmeter reading gives a fair idea of the amount of charge remaining in the cell. However, during a considerable portion of the discharge the voltage remains nearly constant and the voltmeter method therefore cannot be very accurate, for which reason the hydrometer method is preferable. An hydrometer is an instrument for determining the density or specific gravity of a fluid. It consists of a glass tube sealed at both ends, the upper portion being of smaller diameter than the lower portion. The lower end is weighted with shot or mercury so that when the instrument is placed in the fluid to be tested it will assume a vertical position. It is well known that a floating body displaces an amount of liquid equal to its own weight and a body of a given weight naturally will displace more of a light than of a heavy liquid; in other words, it will sink more deeply into the light liquid. The top portion or stem of the hydrometer is graduated and the figure appearing on the instrument at the level of the liquid shows the specific gravity of the latter or its density, according to some conventional scale.

Baumé Scale—Hydrometers are often graduated according to the Baumé scale. The Baumé scale for liquids heavier than water is based upon the following equation:

$$\text{Sp. Gr.} = \frac{145}{145 - \text{Baumé degrees}} \text{ at } 60° \text{ F.}$$

The following table gives the corresponding specific gravities and Baumé degrees:

SECONDARY OR STORAGE BATTERIES.

Specific Gravity Baumé Conversion Table

Bé.	Spec. Grav.	Bé.	Spec. Grav.
0	1.000	18	1.141
1	1.006	19	1.150
2	1.014	20	1.160
3	1.021	21	1.169
4	1.028	22	1.178
5	1.035	23	1.188
6	1.043	24	1.198
7	1.050	25	1.208
8	1.058	26	1.218
9	1.066	27	1.228
10	1.074	28	1.239
11	1.082	29	1.250
12	1.090	30	1.260
13	1.098	31	1.271
14	1.106	32	1.283
15	1.115	33	1.294
16	1.124	34	1.306
17	1.132	35	1.318

FIG. 19.

Hydrometer Syringe—Owing to the fact that storage cells are generally sealed, it is impossible to introduce an hydrometer into them directly, and it has become customary to combine the hydrometer with a syringe by means of which some of the electrolyte is withdrawn from the cell through the opening into which the vent plug screws. As shown in Fig. 19, the hydrometer is inside the glass syringe and can be read through the wall of the latter. After the reading has been taken, the electrolyte is returned to the cell. When water has been added to a cell, no gravity test should be taken before the cell has been charged, as otherwise no correct reading will be obtained.

When a cell is fully charged the specific gravity should be between 1.28 and 1.30. The reduction of the specific gravity by the discharge, of course, depends upon the proportion of electrolyte to active material, but generally the density when the battery has been discharged to a voltage of 1.8 volts is between 1.12 and 1.14. If the densities at complete charge and at the lowest point of charge are known, it is an easy matter to determine the proportion of charge in the battery corresponding to any specific gravity of the electrolyte, since the specific gravity increases and decreases directly with the charge. For instance, suppose that the specific gravity is 1.12 when the cell is discharged as far as desirable and 1.28 when fully charged. Then it is evident that at quarter charge the specific gravity is 1.16; at half charge, 1.20; and at three-quarter charge, 1.24.

Adjusting Level of Electrolyte—The level of electrolyte in the cell gradually drops, due to evaporation, and as the whole of the plates must be covered by electrolyte in order that the full capacity may be available, the level must be corrected from time to time. Nothing but water evaporates and therefore only water (distilled or rain water) should be added. Of course, electrolyte may be lost by spillage, in which case the loss should be made good by adding new electrolyte made up by adding one part by volume of pure sulphuric acid to three parts of distilled water. It is very important to remember in this connection that under no conditions should water be poured into sulphuric acid—always the sulphuric acid into the water. The acid is poured slowly into the water while the solution is stirred with a wooden paddle. When the two are combined a chemical action takes place which generates heat, and if water were poured into the acid the heat production would be apt to be so precipitate that steam would be formed and sulphuric acid be splashed around, which would be dangerous. Since sulphuric acid attacks most of the common metals, the solution must be prepared in glass or earthenware vessels. Before the newly formed solution is poured into the cells, it must be allowed to cool. The best time to add distilled water to cells to make up for loss by evaporation is before a charge.

Charging—In charging a battery great care must be taken to make the proper connections. Usually the terminals of a battery are marked as to which is positive and which negative, and the positive of the battery must be connected to the positive of the charging source and the negative of the battery to the negative of the charging source. Only "continuous" current can be used for charging.

It is in general better to charge at a comparatively slow rate than at a fast rate. Slow charging is more efficient than fast charging and is not injurious to the battery. In a self-contained lighting and starting system the rate of charge, of course, is determined by the design of the generator and its control system, and the rate is generally comparatively low. In charging from an independent source, battery makers usually recommend that charging be started at a rate in amperes equal to one-sixth the capacity of the battery at a 10-hour discharge rate, that this be kept up until the cell registers

SECONDARY OR STORAGE BATTERIES.

2.55 volts and that the charge be then continued at one-half or one-third this rate until either the voltage or the density of the electrolyte ceases to increase. For instance, in the case of a battery rated at 120 ampere-hours for a 10-hour discharge the charge would be begun at 20 amperes and continued at that rate until the voltage were 2.55 per cell and would then be continued at 7 to 10 amperes. As soon as charging is stopped, the voltage drops, and when discharge begins it drops still further. This is the ideal method of charging and would be adopted where large batteries are concerned and the charging facilities permit, but in case of a small battery and especially if several somewhat different ones have to be charged in series at one time, it is best to use such a current that the charge will be completed in, say, 24 hours, at a uniform rate.

Too high a rate of charge or discharge results in buckling of the plates and in shedding of active material. If too high a charging current is sent into the battery, and especially if a battery is overcharged, it will "gas"—that is, a myriad of small gas bubbles will be formed and rise to the surface, giving the electrolyte a milky appearance. The bubbles are formed in the active material where the chemical action takes place and they tend to promote the shedding of the material. Continued overcharges are to be avoided, as they are both wasteful and injurious, but occasional overcharges at slow rates are to be recommended, as they overcome the tendency of the plates to become sulphated.

Sulphating—During each discharge both the positive and negative plates become covered with lead sulphate, but in the normal use of the battery the sulphate is converted during the following charge to lead peroxide on the positive and spongy lead on the negative plate. If, however, the battery is allowed to stand in the discharged state for any length of time, the sulphate on the plates will harden, causing the phenomenon known as sulphating. This will result, when the battery is put in use again, in loss of capacity, buckling, shedding of active material and greater heating of the cells due to increased internal resistance.

Sulphating can be cured by continued overcharging at a rather slow rate. This loosens the sulphate and restores the plates to their normal condition.

Efficiency—A storage cell has a lower effective voltage on discharge than on charge, as may be seen from Fig. 15. Besides, the quantity of electricity in ampere-hours which can be obtained from it is also less than that sent into it in charging. Therefore, there is a certain loss in the battery and its efficiency is always below 100%. It is customary to distinguish between ampere-hour efficiency and watt-hour efficiency, the former being the ratio of output to input in ampere-hours and the latter the ratio of output to input in watt-hours. The former is always the greater but the latter is the real efficiency in a commercial sense.

Loss of Capacity—One of the most common troubles with storage batteries is loss of capacity. This may be due to one or another of three general causes, viz.: weak electrolyte, sulphating and shedding of active material. The gravity of the electrolyte should occasionally be tested at the completion of a charge, and if it is not at least 1.28, corrected by adding

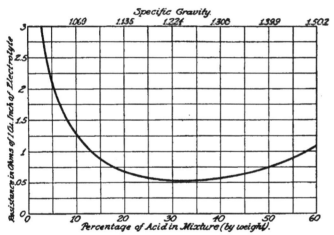

Fig. 20—Variation of Electrical Resistance with Strength of Electrolyte.

extra strong electrolyte. Weak electrolyte has a greater electric resistance than electrolyte of normal strength, as may be seen from Fig. 20, and as a result weakness of the electrolyte reduces the efficiency of the cell. During the whole life of a cell the active material is gradually shed from the plates and accumulates as sediment at the bottom of the cell. It must not come nearer than about ¼ inch to the bottom of the plates, as there would then be danger of the plates being short-circuited by any additional active material that might be shed.* Formerly it was customary after a battery had been used for a certain length of time to remove the plates from the jars and wash the latter out, but at the

SECONDARY OR STORAGE BATTERIES. 31

present time such high bridges are customarily provided that when the mud space underneath the plates is substantially filled with sediment, the cell has lost so much of its capacity that it is best to renew the plates. This has done away with washing the cells.

If a car equipped with a storage battery for ignition is to be laid up for a period of time, it must be given a charge at least once every two months and also just prior to putting it into use again. As regards the installation and care of batteries for lighting and starting I cannot do better than to quote the following instructions issued by the Society of Automotive Engineers:

S. A. E. Instructions

1. Batteries must be properly installed.

Keep battery securely fastened in place.

Battery must be accessible to facilitate regular adding of water to, and occasional testing of, solution. Battery compartment must be ventilated and drained, must keep out water, oil and dirt and must not afford opportunity for anything to be laid on top of battery. Battery should have free air spaces on all sides, should rest on cleats rather than on a solid bottom and holding devices should grip case or case handles. A cover, cleat or bar pressing down on the cells or terminals must not be used.

2. Keep battery and interior of battery compartment wiped clean and dry.

Do not permit an open flame near the battery.

Keep all small articles, especially of metal, out of and away from the battery. Keep terminals and connections coated with vaseline or grease. If solution has slopped or spilled, wipe off with waste wet with ammonia water.

3. Pure water must be added to all cells regularly and at sufficiently frequent intervals to keep the solution at the proper height.

The proper height for the solution is usually given on the instruction or name-plate on the battery. In all cases the solution must cover the battery plates.

The frequency with which water must be added depends largely upon the battery, the system with which it is used and the condition of operation. Once every two weeks is recommended as good practice in cool weather, once every week in hot weather.

Plugs must be removed to add water; then replaced and screwed home after filling.

Do not use acid or electrolyte, only pure water.

Do not use any water known to contain even small quan-

32 SECONDARY OR STORAGE BATTERIES.

tities of salts of any kind. Distilled water, melted artificial ice or fresh rain water are recommended.

Use only a clean, non-metallic vessel.

Add water regularly, although the battery may seem to work all right without it.

4. The best way to ascertain the condition of the battery is to test the specific gravity (density) of the solution in each cell with a hydrometer.

This should be done regularly.

A convenient time is when adding water, but the reading should be taken before, rather than after, adding the water.

A reliable specific gravity test cannot be made after adding water and before it has been mixed by charging the battery or by running the car.

The cut (see Fig. 18) illustrates a common and convenient form of hydrometer syringe to test the specific gravity of the electrolyte. To take a reading, insert the end of the rubber tube in the cell. Squeeze and then slowly release the rubber bulb, drawing up electrolyte from the cell until the hydrometer floats. The reading on the graduated stem of the hydrometer at the point where it emerges from the solution is the specific gravity of the electrolyte. After testing, the electrolyte must always be returned to the cell from which it was drawn.

The gravity reading is expressed in "points;" thus the difference between 1,250 and 1,275 is 25 points.

5. When all cells are in good order the gravity will test about the same (within 25 points) in all.

Gravity above 1,200 indicates battery more than half charged.

Gravity below 1,200 but above 1,150 indicates battery less than half charged.

When battery is found to be half discharged, use lamps sparingly until by charging the battery the gravity is restored to at least 1,200. See Section 8.

Gravity below 1,150 indicates battery completely discharged or "run down."

A run-down battery should be given a full charge at once. See Sections 7 and 8.

A run-down battery is always the result of lack of charge or waste of current. If, after having fully charged, the battery soon runs down again, there is trouble somewhere else in the system, which should be located and corrected.

Putting acid or electrolyte into the cells to bring up specific gravity can do no good and may do great harm. Acid or elec-

SECONDARY OR STORAGE BATTERIES. 33

trolyte should never be put into the battery except by an experienced battery man.

6. Gravity in one cell markedly lower than in the others, especially if successive readings show the difference to be increasing, indicates that the cell is not in good order.

If the cell also regularly requires more water than the others, a leaky jar is indicated.

Even a slow leak will rob a cell of all of its electrolyte in time, and a leaky jar should be immediately replaced with a good one.

If there is no leak and if the gravity is or becomes 50 to 75 points below that in the other cells, a partial short-circuit or other trouble within the cell is indicated.

A partial short-circuit may, if neglected, seriously injure the battery and should receive the prompt attention of a good battery repair man.

7. A battery charge is complete when, with charging current flowing at the rate given on the instruction-plate on the battery, all cells are gassing (bubbling) freely and evenly and the gravity of all cells has shown no further rise during one hour.

The gravity of the solution in cells fully charged as above is 1,275 to 1,300.

8. The best results in both starting and lighting service will be obtained when the system is so designed and adjusted that the battery is normally kept well charged, but without excessive overcharging.

If, for any reason, an extra charge to maximum specific gravity is needed, it may be accomplished by running the engine idle, or by using direct current from an outside source.

In charging from an outside source use *direct* current only. Limit the current to the proper rate in amperes by connecting a suitable resistance in series with the battery. Incandescent lamps are convenient for this purpose.

Connect the positive battery terminal (painted red, or marked POS or P or +) to the positive charging wire and negative to negative. If reversed, serious injury may result. Test charging wires for positive and negative with a voltmeter or by dipping the ends in a glass of water containing a few drops of electrolyte, when bubbles will form on the negative wire.

9. A battery which is to stand idle should first be fully charged. See Sections 7 and 8.

A battery not in active service may be kept in condition for use by giving it a freshening charge at least once every two months, but should preferably also be given a thorough

34 SECONDARY OR STORAGE BATTERIES.

charge, after an idle period, before it is replaced in service.

A battery which has stood idle for more than two months should be charged at one-half normal rate to maximum gravity before being replaced in service.

It is not wise to permit a battery to stand for more than six months without charging.

Disconnect the leads from a battery that is not in service so that it may not lose through any slight leak in car wiring.

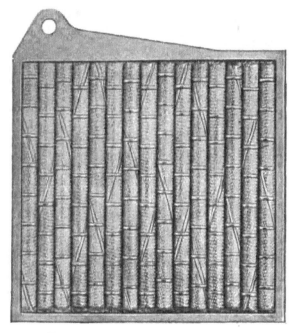

FIG. 21—POSITIVE PLATE OF EDISON CELL.

The Edison Nickel-Alkaline Battery.—A storage battery employing nickel and iron salts for the active materials and an alkaline solution for the electrolyte has been invented by Thomas A. Edison. It is used extensively for the propulsion of electric commercial vehicles, industrial trucks, etc., and has also been adapted for ignition and lighting purposes, and even for starting, though the high discharge rate entails a serious drop in voltage. All the structural elements of the Edison cell—the jar or can and the positive and negative plates—are made of sheet steel, nickel-plated. The positive plate (Fig. 21) consists of tubes of perforated sheet steel, arranged vertically and fitted into a sheet steel grid. The tubes are filled with the active material, nickel-hydrate, interspersed with thin layers of nickel flakes to increase the conductivity. They are spirally wound and the edges are double seamed. Nickel-plated steel rings are spaced along the tube on the outside to prevent it from expanding.

Iron oxide mixed with mercury oxide serves as the nega-

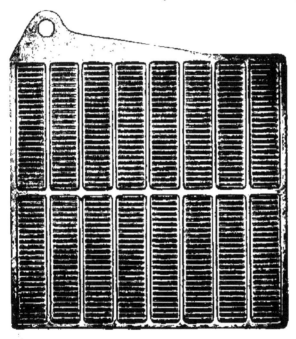

FIG. 22—NEGATIVE PLATE OF EDISON CELL.

tive active material and is contained in rectangular pockets made of perforated nickel-plated steel ribbon (see Fig. 22). After the pockets have been filled with the active material, the plate is re-formed in a press which slightly corrugates the pockets and brings the sheet metal into more intimate contact with the active material, thus decreasing the internal resistance. The plates are assembled into positive and negative groups respectively by means of threaded steel rods

36   SECONDARY OR STORAGE BATTERIES.

passing through holes in one corner of the plates and steel spacing washers. To the middle of the rod is secured a terminal post. The negative group always contains one more plate than the positive, for the same reason as in the case of the lead storage battery. In order to keep positive

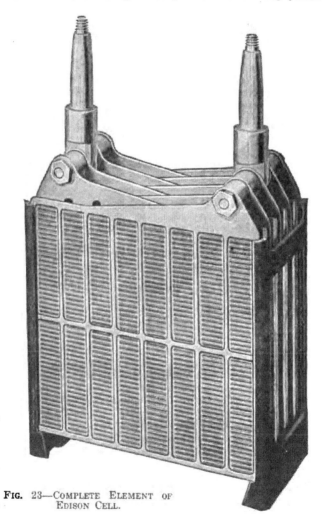

FIG. 23—COMPLETE ELEMENT OF EDISON CELL.

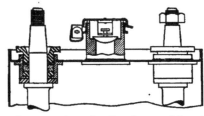

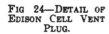
Fig 24—Detail of Edison Cell Vent Plug.

and negative plates apart separators in the form of hard rubber rods are inserted between them. The can is made of thin sheet steel corrugated for stiffness in the larger sizes, with joints made by the oxy-acetylene welding process. Into this the element is inserted; it stands on hard rubber bridges on the bottom of the can and is kept out of contact with the sides of the latter by spacers of hard rubber over its ends. The cover for the can is also of sheet steel and contains gland fittings through which the electrodes pass, the gland nuts, washers and bushings being of insulating material A combined filling aperture and vent plug is secured to the center of the cover. Details of the insulation of the electrodes and of the vent plug are given in Fig. 24.

The electrolyte consists of a 21 per cent solution of potassium hydrate in distilled water, to which is added a small amount of lithium hydrate.

For 6-volt ignition and lighting service, 5-cell batteries are used, which are put up in wooden trays contained in a steel box that bolts to the battery support. The average voltage during discharge is 1.2 per cell, or 6 for the five cells. The voltage is not quite so constant as in the case of the lead battery. At the beginning of the discharge it stands at 1.46 volts, but it quickly drops to 1.3, and after about one-half the charge has been taken out, it is down to 1.2. For a given capacity, the Edison cell is lighter than the lead cell, but somewhat bulkier. This comparison applies to the cell proper and not to the complete battery with its wooden box or tray, which latter adds considerably to the bulk but not very much to the weight. At present, Edison batteries are made for lighting and ignition in two sizes, and the chief data of these are as follows:

No. of Cells	Volts	Total Weight	Dimensions
5	6	25 lbs.	8⅞"x6¾"x10 "
5	6	40 lbs.	8⅞"x6¾"x15½"

The battery can be completely charged in 7 hours. Constant current charging is preferable, but constant potential charging is permissible, and if this method is used it is ad-

SECONDARY OR STORAGE BATTERIES.

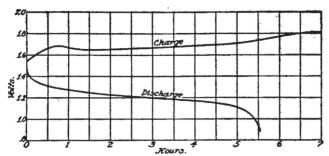

FIG. 25—CHARGE AND DISCHARGE CURVES OF EDISON CELL.

visable to start at a rate 50% greater than the normal rate, in which case the average rate will be equal to the normal. Over-charging at the normal rate has no harmful effect, and it is recommended to give the battery a twelve-hour charge once every sixty days or when the electrolyte is renewed. Of course, when the battery forms part of a self-contained electric system on the car, all that is necessary is to see that the car is so driven and the charge regulator (if one is provided) so set that the charge never gets low.

The electrolyte must be kept at a level so as to completely cover the plates, and any loss by evaporation must be made up by adding distilled water. No injury is suffered by the plates while the cells stand in a discharged condition.

Among the precautions to be observed in caring for Edison batteries are the following: The cells externally must be kept clean and dry, because the container or can is made of conducting material. The filler caps must be kept closed, except when replenishing the electrolyte. The cells must be kept filled with electrolyte to the proper level. No tools or other metal pieces must be laid on top of the battery, and under no conditions must acid be poured into the cells.

Edison batteries are more expensive than lead batteries of equal capacity, but are claimed to have a much longer life.

CHAPTER IV

Magnetism

When a freely suspended bar of steel or iron is brought into the vicinity of a wire carrying an electric current the bar tends to assume a position in a plane perpendicular to the wire. This is due to the fact that the electric current flowing through the wire creates a magnetic field of force in the vicinity of the wire, and the magnetic force acts on the steel bar or rod. So-called lines of magnetic force encircle the wire, as can very nicely be demonstrated by arranging the wire vertically, threading it through a sheet of white paraffined paper and sprinkling iron filings on the paper while the current is flowing. The filings will arrange themselves with their longer dimensions transverse to the wire, in such a way as to form a series of circles around it. If the wire carrying the current is of considerable length, the strength of the magnetic force at any point in the vicinity of the wire is directly proportional to the current strength and inversely proportional to the distance from the wire.

The Helix or Coil.—To produce a more intense magnetic field with a limited current, the wire must be wound into a helix (Fig. 26). It is readily seen that in that case all of the magnetic lines which would encircle the wire of the helix, if stretched out, within a distance equal to the diameter of the helix, will be within the latter. If, when the helix is looked at from one end, the current flows through it in a clockwise direction, then the direction of the lines of force is away from the observer.

Such a helix has several peculiar properties. When freely suspended it will turn till its axis extends north and south. If a bar of iron or steel is brought near it, there will be mutual attraction between the two. In fact, the tendency of the current is to draw the metal piece into the center of the helix When a bar of iron or steel is thus introduced into a helix through which an electric current flows, the magnetic effect becomes greatly augmented, owing to the fact that iron and steel, the so-called magnetic metals, offer circuits. This law of the magnetic circuit is as follows:

$$\text{Flux} = \frac{\text{Magneto motive force}}{\text{Reluctance}}$$

The practical unit of magneto motive force is the ampere-turn, that is, a current of one ampere flowing through one convolution of the helix or coil. In any particular case the number of ampere turns is obtained by multiplying the number of turns in a coil by the amperes flowing through it. The reluctance of a magnetic circuit is proportional to its length and inversely proportional to its cross-sectional area and its permeability. The permeability of air, which is con-

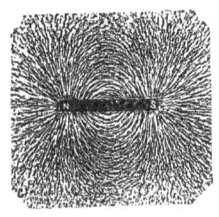

FIG. 29—FIELD OF FORCE OF A BAR MAGNET.

stant, is taken as unity. The permeability of the magnetic metals is not constant, but decreases as the flux density increases.

In general

$$\text{Reluctance} = \frac{\text{Length}}{\text{Cross Section} \times \text{Permeability}}$$

and if the length and section in this equation are given in centimeters and square centimeters, respectively, then the reluctance is obtained in international units known as oersteds. However, all calculations can readily be made by using inch units, as will be shown further on.

Fig. 30 is a diagram showing the relation between the flux density and the magneto motive force, in wrought iron and cast iron, respectively. It should be pointed out that this relation depends upon the composition and mechanical state of the material, and the curves represent average values

MAGNETISM.

only. The density of magnetization B, or lines of force per square inch of cross section, is plotted along the vertical axis, and the magneto motive force required, in ampere-turns per inch in length, is plotted along the horizontal axis. It will be noticed that in the case of each of these metals, as the magneto motive force attains a considerable value, the flux density increases less rapidly. The point where the curve shows a decided bend is known as the saturation point.

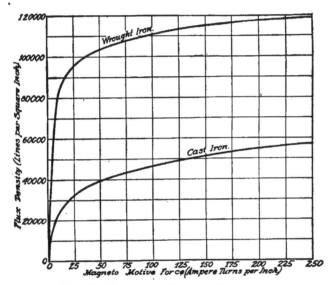

FIG. 30—CURVES OF MAGNETIC INDUCTION.

To produce a magnetic intensity of B lines per square inch in an air gap l inches long requires

$$\frac{B}{3.21} \text{ ampere-turns}$$

and to produce the same intensity in magnetic material of permeability μ requires

$$\frac{B}{3.21\mu} \text{ ampere-turns.}$$

The permeability μ of a magnetic material is the ratio of the magneto motive force required to produce a certain flux density in air to the magneto motive force required to produce the same flux density (lines per square inch) in the material under consideration. This factor varies with the flux density, and its values for different densities B can be obtained from Fig. 31. It will be noticed that at the point where saturation begins the permeability has a value of about 1500 for wrought iron and about 500 for cast iron. This makes it plain why that part of any magnetic circuit passing through air should be as short as possible, as it takes hundreds of times the magneto motive force to produce a certain flux density in air as in magnetic material, if length and cross sectional area are the same.

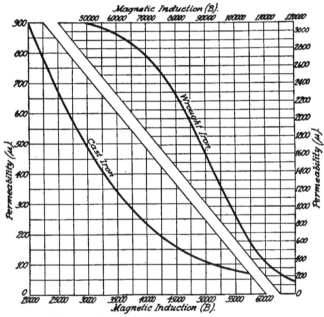

FIG. 31—CURVES OF MAGNETIC PERMEABILITY.

Permanent Magnets.—When a piece of hardened steel is subjected to a magnetic force it acquires permanent magnetic properties, as already pointed out. When the outside magnetizing force ceases, the magnetic strength or flux decreases. In order that the permanent magnet may be comparatively strong and keep its strength well, the ratio of its length to its sectional area should be large, and the length of its air gap should be small as compared with the length of the magnet itself. In fact, the magnet will keep

MAGNETISM.

its strength best if its magnetic circuit is completely made up of magnetic material, and for this reason it is usual to provide horseshoe magnets with an armature or keeper A, Fig. 32, when they are not in use.

If a bar magnet is broken in two, each half is a complete magnet, that is, it possesses both a north pole and a south pole. This shows that the coercive force which causes the field of force of a permanent magnet (and which corresponds to the magneto motive force in an electro magnet) is distributed over the whole length of the magnet. In fact, other things being equal, the coercive force varies directly as the length of the bar, and is independent of its cross section. This explains the advantage of relatively long magnets, for if we double the length of the magnet the average

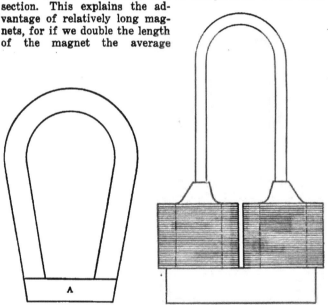

FIG. 32—HORSESHOE MAGNET.   FIG. 33—MAGNETIZING A HORSESHOE MAGNET.

length of the air path of the lines of force will not quite be doubled, and as the air path constitutes the chief reluctance the flux will be greater. In other words, the poles will be stronger.

When comparatively large magnets are desired they are usually built up from a number of equal or nearly equal sections. Probably the chief reason for this is that if the cross section of the individual magnet is small the latter can more readily be hardened uniformly all through, and great hardness is one of the conditions of a high degree of permanence.

Magnet Material.—The best results are obtained from permanent magnets made of tungsten steel containing 5 to 6 per cent of tungsten and from 0.60 to 0.75 per cent of carbon. Equally high magnetic permeability is obtained with ordinary carbon steel, but the tungsten increases the degree of permanence. The two requirements in magnet steel, viz., high permeability and high retentivity, conflict with each other, as soft steel has the highest permeability and hard steel the highest retentivity. For this reason a compromise is made and a medium carbon steel used (less carbon than in

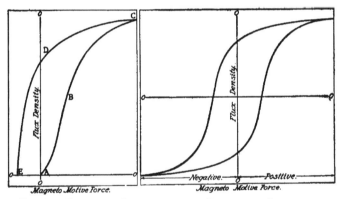

Fig. 34—Curves of Increasing and Decreasing Induction.
Fig. 35—Hysteresis Curve.

spring steel). The magnets are forged to shape at as low a temperature as possible, are then raised to a temperature of 1500-1600 degrees Fahr. and quenched in cold water or brine. In order to insure uniform hardness they must be absolutely free from scale during the quenching operation.

There are two methods of magnetizing or charging horseshoe magnets. One consists in inserting the poles into solenoids or coils through which a powerful current is flowing, the other in applying them to the poles of a strong electro magnet (Fig. 33). According to modern theory, the effect of magnetization is to turn all of the molecules of the steel, which ordinarily lie in all possible directions, in a certain direction, and to facilitate this turning motion it is the practice of some magnet makers to vibrate or tap the magnets while they are under the magnetizing influence. Another expedient which is believed to favorably influence the remanent magnetism is to decrease the magnetizing force gradually.

After the magnetizing influence has ceased and the magnetic flux in consequence decreased, the magnet is in a somewhat unstable state. There are three general causes tending to further reduce its strength, viz., reverse magnetizing forces (which may be due to electric current flowing close to it), changes in temperature, and shock. For many purposes it is far more desirable to have a magnet whose strength will remain substantially constant throughout a long period than a magnet of the greatest possible initial strength, and in such cases an aging process is applied to the magnets after they have been magnetized. Probably the best aging process consists in applying to them all of the three demagnetizing influences above enumerated. That is, they may be tapped or tumbled and subjected to the action of live steam and to a comparatively small counter magneto motive force. Such an elaborate aging process is to be recommended for magnets for measuring instruments whose permanency determines the dependability of the indications, but for ordinary purposes any one of the three processes alone will serve.

Lifting Power of Magnets.—The force with which a horseshoe magnet holds its armature may be calculated by means of the equation

$$F = \frac{2.7 \times B^2 \times s}{100,000,000} \text{ pounds}$$

where B represents the magnetic induction in lines per square inch and s the cross section in square inches. Electro magnets are often used for lifting purposes, and will readily lift 125 pounds per square inch of section with a sufficient factor of safety.

Hysteresis.—If a bar of iron is magnetized by a gradually increasing current, its magnetic induction will increase in accordance with the curve shown in Fig. 30. This same curve is reproduced on a smaller scale in Fig. 34, being represented by ABC. If after a certain induction has been attained the current is gradually decreased, it will be found that the iron is more highly magnetized while the current has a certain value during the decrease than during the increase. The relation between the magnetizing force and the induction is represented in the figure by the curve CDE. This effect is due to the retentivity of the iron and steel, in consequence of which, as we have already seen, when the

MAGNETISM.

magnetizing current ceases the steel retains a certain amount of magnetism. It requires a slight counter magneto motive force EA applied to the steel to deprive it of this magnetism. This value, EA, is a measure of the coercive force of the magnet. In certain practical applications of electro-magnets the magnetizing force is rapidly reversed in direction. The magnetization and demagnetization then follow the curve shown in Fig. 35. The fact that the iron does not spontaneously lose its magnetism when the magnetizing force is removed, but actually has to be demagnetized by a counter magneto motive force, results in a certain loss known as hysteresis loss. The area included between the two curves in Fig. 35 is a measure of this loss, and the diagram is known as a hysteresis diagram.

CHAPTER V

Electromagnetic Induction

It was stated in the preceding chapter that a wire carrying a current is encircled by lines of magnetic force. The density of these lines is greatest near the wire and decreases with an increase in the distance from the wire. If now another wire be stretched parallel with the first and close to it, then the majority of the magnetic lines of force encircling the first wire will also encircle the second, which latter is not supposed to be connected to any source of current. There will be no effect on the second wire as long as the current in the first remains constant, but as soon as there is any change in the value of this current an electromotive force will be induced in the second wire. Suppose that the current in the first wire is increased. Then the density of the magnetic lines of force in its vicinity is increased. The new lines may be supposed to emanate from the wire carrying the current and to expand until a state of equilibrium is attained. In doing so many of them will link the second wire and it is this linking of lines of magnetic force with a conductor that produces the electromotive force. When the current in the first wire decreases in value and in consequence the number of lines encircling the second wire is reduced, the result is exactly the same, except that the induced electromotive force in this case is in the opposite direction. In both cases the lines of force are cut by the conductor.

In order to strengthen this effect we again substitute coils for straight wires. Suppose that two coils of wire be located coaxially and close together, as shown in Fig. 36. Let one coil be in circuit with a source of current and let a current indicator or galvanometer be included in the circuit of the second coil. Then when the first circuit, called the primary circuit, is either made or broken, the galvanometer will indicate a current impulse in the other circuit, which is known as the secondary circuit. On the closing of the first circuit a certain number of lines of force are linked with each of the convolutions of the secondary coil, and on the opening of the primary circuit they are "unlinked."

Law of Electro-Magnetic Induction.—This phenomenon of the action of a current-carrying conductor upon another conductor is known as electro-magnetic induction. It was discovered by Faraday, whose experimental results may be summed up as follows: "If the number of magnetic lines of force through any circuit be varied by any means, an electromotive force is set up in that circuit proportional at any in-

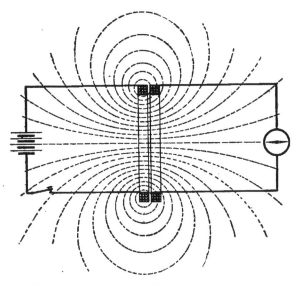

FIG. 36—ILLUSTRATING INDUCTIVE EFFECT OF ONE COIL UPON ANOTHER DUE TO STARTING AND STOPPING OF CURRENT IN THE FORMER.

stant to the rate of change in the number of lines of force at that instant."

There are, of course, various means of varying the magnetic induction through a circuit. One method has already been described, viz., by starting and stopping a current in an adjacent circuit. Another method consists in introducing a magnet into a coil of wire or withdrawing it therefrom, as illustrated in Fig. 37. A third method consists in rotating a coil around a given axis in a magnetic field so that the plane of the coil is soon parallel and soon perpendicular to the lines of force. This is illustrated in Fig. 38. The arrows in

ELECTROMAGNETIC INDUCTION.

the latter figure show the directions, respectively, of the lines of force, the rotation and the induced electromotive force.

The electromotive force in each case is proportional to the rate of cutting the lines of force. If 100,000,000 lines are cut by one conductor in one second, then the electromotive force induced in that conductor is 1 volt. In the case of a coil of many turns the number of lines introduced into or withdrawn from the coil per second must be multiplied by the

FIG. 37—ILLUSTRATING INDUCTIVE EFFECT IN A COIL DUE TO THE INTRODUCTION OF A PERMANENT MAGNET.

number of turns in the coil to get the induced electromotive force.

Direction of Induced Currents.—In the case of two parallel wires or two adjacent coils, if the current in the first increases the induced electromotive force in the second is opposite in direction to that in the first, but when the current in the first decreases the induced electromotive force in the second has the same direction as the current in the first.

Self-Induction.—Faraday's rule says that if the number of lines of force through any circuit be varied by any means an electromotive force will be set up in that circuit. One means

52   ELECTROMAGNETIC INDUCTION.

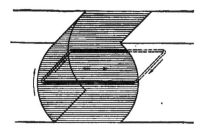

Fig. 38—Induction due to Rotation of a Coil in the Magnetic Field.

evidently consists in sending a variable current through the circuit. This will cause a variation in the flux through the coil and induce in it what is called an electromotive force of self-induction. As long as the current increases the electromotive force of self-induction is opposed to it, and tends to prevent its further increase, but when the current decreases, the electromotive force of self-induction is in line with the current and tends to prevent its further decrease. One result of self-induction is that when a circuit including a source of current is closed the current does not instantly attain its maximum value. Supposing the electromotive force to be constant, the growth of the current is represented by the curve in Fig. 39. Theoretically it takes an infinite time for the current to reach its maximum value, the quotient of the electromotive force by the resistance, which is represented by the straight horizontal line above the curve, but in practical cases it attains a value very near the maximum in an exceedingly short time—a small fraction of a second. The time which it takes for a current to attain a value equal to 63.2 per cent. of its maximum is known as the time constant of the circuit. This is the greater

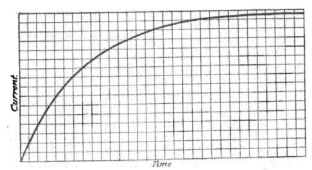

Fig. 39—Growth of Current after Circuit is Closed.

ELECTROMAGNETIC INDUCTION. 53

the higher the self-inductance. It is possible to make a coil without self-inductance by doubling up a wire and winding it on a spool that way, in which case there will be an equal number of positive and negative turns, and consequently no ampere-turns and no magnetic flux. The self-inductance of a circuit is equal to the number of lines of force set up by a current of 1 ampere. If the circuit consists of a given length of wire in a single loop the self-inductance will be comparitively small. It can be increased by forming the wire into a coil, further increased by putting an iron core into the coil, and still further by providing the coil with a magnetic circuit wholly of iron.

The effect of self-inductance may be clearly demonstrated as follows: Take a battery of six dry cells in series. Connect a wire to one of the terminals of the battery and brush it against the other. A hardly perceptible spark will be produced as the circuit is broken. Now connect a so-called choke coil in circuit. This consists of a core of fine iron wire over which is wound a coil of insulated copper wire, of, say, No. 14 B. & S. gauge. When the circuit is again closed and broken by brushing the free end of the wire over a terminal, a very powerful spark is obtained. This shows that the electromotive force of self-induction in this case is very much greater than the electromotive force of the battery.

Sine Wave.—Suppose that a single coil of wire, of rectangular shape, be rotated around a central axis in a magnetic field of uniform strength, as indicated in Fig. 38. The end portions of the coil move in a plane parallel to the lines of force and therefore do not cut any. The two sides of the coil, however, cut the lines more or less rapidly, according to their momentary position, and it will be seen that the electro motive forces induced in the two lengths of wire are in series and add together. Evidently, the lines are being cut at the highest rate when the coil is in the horizontal position (as shown), and this position corresponds to the maximum induced electromotive force. As the coil passes by this position the electromotive force decreases, at first slowly and then more rapidly, and when the vertical position is reached the electromotive force is nil, because the sides of the coil move momentarily parallel with the lines of force. The next instant the direction of the induced electromotive force is reversed. The waves of electromotive force induced in such a coil may be represented by the curve shown in Fig. 40, which is known as a sine curve. If the circuit of the coil be closed then a current will flow in it which rises and falls and changes in direction substantially like

54 ELECTROMAGNETIC INDUCTION.

the electromotive force. Such a current is known as an alternating current.

In stating the value of the current or electromotive force in an alternating current circuit the average value during a half cycle is the proper figure to use. In the case of a true sine curve the average value is 63.7 per cent. of the maximum, and is represented by the line a—a, in Fig. 40. However, in commercial work alternating currents do not necessarily follow the sine curve, and the proportion between maximum and mean value therefore may be slightly different.

A rule for the direction of electromotive forces induced by the motion of a conductor in a magnetic field is as follows:

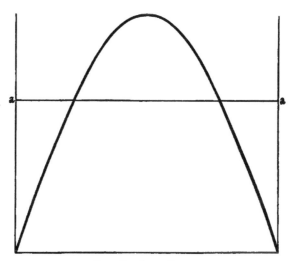

FIG. 40—SINE CURVE.

Extend the thumb, middle finger and index finger of the right hand at right angles to each other, as shown in Fig. 41. If the index finger indicates the direction of the lines of force and the thumb the direction of motion, then the middle finger will indicate the direction of the induced electromotive force.

Effects of Self-Induction.—Referring to Fig. 42, we will suppose that a current represented by the sine curve C flows in a circuit having a certain self-inductance (see sketch in upper right-hand corner). Then the change in value and

ELECTROMAGNETIC INDUCTION. 55

direction of the current will set up an electromotive force of self-inductance of which we know that it is co-directional with the current when the latter decreases in value and opposed to the current when the same increases in value. When the current does not change in value there is no change in the number of magnetic lines, and hence the electromotive force of self-inductance is nil. This is the case when the current has its maximum positive and negative values. At these points, therefore, the electromotive force of self-inductance is zero. These considerations enable us to draw a curve of the electromotive force of self-inductance, and this is represented by the dotted curve S, Fig. 42. Now the

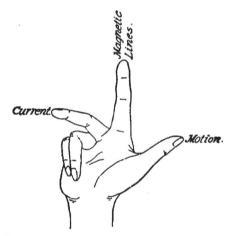

FIG. 41—SKETCH ILLUSTRATING A METHOD OF DETERMINING THE DIRECTION OF INDUCED ELECTROMOTIVE FORCE.

source of current has to overcome both the resistance of the circuit (sometimes called the ohmic resistance) and also the electromotive force of self-inductance, which is at least partly opposed to it. Since the electromotive force required to overcome the ohmic resistance is at every moment directly proportional to the current, we may assume it to be also represented by the curve C. The electromotive force to overcome inductance is represented by curves S_1, which is at every point equal and opposite to S, and the impressed electromotive force (that of the source of current) is the resultant of the two, which is represented by curve R.

Current Lag.—It will be observed that the maximum current does not occur at the same time as the maximum electro motive force, but a moment later. Also, the current changes direction a moment later than the electromotive force. The current is said to be out of phase with the electromotive force or to lag behind it, and this is due to the inductance of the circuit. This lag may be expressed in degrees.

One of the most peculiar characteristics of an alternating

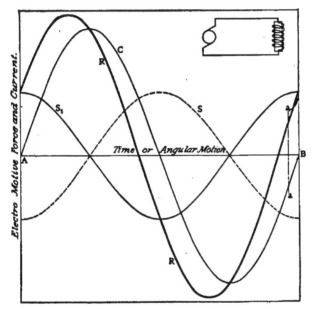

FIG. 42—ELECTROMOTIVE FORCE AND CURRENT WAVES IN AN INDUCTIVE CIRCUIT.

C—Current; also Electromotive Force Necessary to Overcome Ohmic Resistance.
S—Electromotive Force of Self-Induction.
S_1—Electromotive Force Necessary to Overcome Self-Induction.
R—Total Applied Electromotive Force, or Resultant of C and S_1.

current circuit is shown by Fig. 42. At certain periods, just after the electromotive force has changed its direction, the current flows in a direction opposite to the electromotive force. This is true at the point represented by the line a—a, for instance.

If the circuit in the case represented by Fig. 42 had only ohmic resistance, then the current flowing in it would be represented by curve R, instead of by curve C. Therefore, the two chief effects of self-inductance in a circuit are to reduce the value of the current flowing and to cause it to lag behind the electromotive force in phase. In the particular case shown, the effects are not very great, but the current may be reduced to a small fraction of what it would be if there were ohmic resistance in circuit only, by making the self-inductance large in comparison with the ohmic resistance.

Electrical Capacity.—An electric condenser consists of two surfaces of conducting material separated by a layer of insulating material. Such a condenser forms a reservoir for the storage of electrical energy. The capacity of a condenser varies directly as the area of the opposing surfaces and as a factor known as the specific inductive capacity of the insulating material, and inversely as the distance between the surfaces. The specific inductive capacity of air is taken as

FIG. 43—DIAGRAM OF A CONDENSER.

unity, and the s. i. c.'s of the insulating materials most commonly used in condensers are as follows: Paraffin, 2—2.5; mica, 5; glass, 6.5—10.

The quantity of electricity which a condenser will hold is equal to the product of its capacity and the voltage impressed upon it. This may perhaps be most readily understood by considering the electricity to be analogous to a compressible fluid, such as gas. The amount of gas a reservoir will hold depends upon the cubical contents of the reservoir and upon the pressure of the gas. The actual construction of condensers for commercial purposes will be taken up later, and it will suffice here to state that it is customary to represent a condenser by the diagram shown in Fig. 43.

Capacity Advances Phase of Current.—Capacity in an electric circuit neutralizes more or less the effect of self-induction. We have seen that self-induction causes a current to lag in phase. A capacity in parallel with the self-inductance will reduce this effect. Also, self-inductance causes a powerful spark to be produced when a current carrying circuit is broken. Capacity reduces the size of the spark and may eliminate it completely. How these results are accomplished may be explained as follows:

58   ELECTROMAGNETIC INDUCTION.

The capacity, electromotive force and quantity of charge in a condenser are connected by the equation:

Charge = capacity × E. M. F.

Since the capacity is constant, it may be seen that any variations in the amount of charge would be accompanied by a corresponding or proportional variation in the electromo-

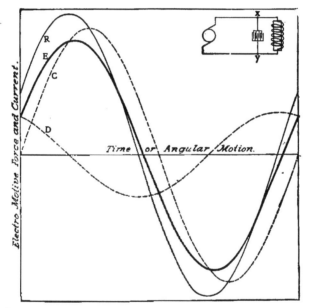

FIG. 44—ELECTROMOTIVE FORCE AND CURRENT WAVES IN A CIRCUIT CONTAINING SELF-INDUCTION AND CAPACITY IN PARALLEL.

R—Electromotive Force Between X and Y.
D—Current Flowing Into Condenser.
C—Current Flowing Through Inductive Circuit.
E—Total Current in Circuit or Resultant of D and C.

tive force. But the rate of increase of the charge is the current. Hence we may write:

Current (flowing into condenser) = Capacity × Rate of Change of Electromotive Force.

We will now assume that a condenser is connected in parallel with a portion of a circuit having both ohmic resistance and self-inductance, as shown in Fig. 44, upper right-hand corner. This may be the same circuit as that shown in Fig. 42, except that the condenser is added. In Fig. 44 the curves of electromotive force and current are reproduced from Fig. 42. The current here represented (curve C) is that flowing through the inductive circuit, and the electromotive force (curve R) that active between points X and Y. This same electromotive force is impressed upon the condenser. We found that the current flowing into the condenser is at all times proportional to the rate of change of the impressed electromotive force. When the electromotive force increases the current is positive (above the line), when the electromotive force decreases the current is negative, when the electromotive force is at either the positive or negative maximum the current is nil, and when the electromotive force passes through the zero points the current is a maximum. These considerations enable us to draw the curve D of current flowing into the condenser, which, it will be seen, is 90 degrees in advance of the impressed electromotive

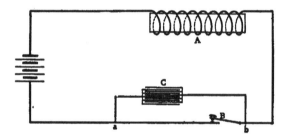

FIG. 45—CONDENSER SHUNTED ACROSS CIRCUIT BREAKER.

force. We now add the two current curves, C and D, together, and obtain the curve E, which represents the combined current flowing in the circuit. It will be seen that this is very nearly in phase with the electromotive force represented by curve R. Hence the capacity neutralizes the phase distorting effect of self-inductance.

Absorption of Extra Current by Condenser—In Fig. 45 is shown a circuit including a choke coil at A and a device B for breaking the circuit. It has been already pointed out that when a circuit including an inductive resistance like A is broken, a powerful spark is produced at the break, owing to the electromotive force of self-induction, which usually far exceeds the electromotive force normally active in the circuit. Another way of looking at the matter is as follows: The magnetic field of the choke coil A represents a certain store of energy, and as the circuit is broken and the current decreases this energy is thrown into the circuit in the form of an extra current, and is dissipated in heat, owing to the resistance encountered by the extra current. The amount of energy stored up in a circuit is proportional to the coefficient of self-induction and to the square of the current flowing previous to the break.

Now, suppose a condenser C to be connected in shunt across the interrupter. As long as the interrupter is closed there is no potential difference between opposite terminals of the condenser (points a and b), and it contains no charge. When the interrupter opens the circuit the potential difference between points a and b quickly increases, and in accordance with the principles explained in the foregoing, a

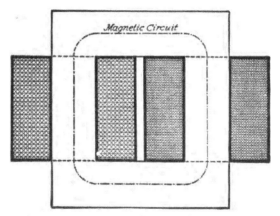

FIG. 46—ALTERNATING CURRENT TRANSFORMER.

current directly proportional to the increase in this potential difference or electromotive force will flow into the condenser and be absorbed there. If the condenser is of the proper capacity it will absorb all of the extra current and there will be no spark at the contact points when the circuit is broken.

The power represented by an alternating current not in phase with its electromotive force is not equal to the product of the current into the electromotive force, but is less, depending upon the angle of lag. If the angle of lag is large a heavy current may flow in a circuit without much energy being expended.

Mutual Induction.—When there is an inductive effect of

ELECTROMAGNETIC INDUCTION.

one coil upon another, as in the case illustrated in Fig. 36, we have mutual induction. If the two coils are so arranged relative to each other that practically all of the lines of force which pass through one must also pass through the other—that is, when there is little magnetic leakage—the electromotive force induced in the secondary coil is to that applied to the primary coil substantially as the number of turns in the secondary coil is to the number of turns in the primary coil. Supposing the secondary to have the greater number of turns, the ratio of the secondary electromotive force to the primary electromotive force will not be quite as great as the ratio of secondary turns to primary turns, because a few of the magnetic lines of force passing through the primary coil will leak past the secondary coil, and also because there is some loss in primary voltage due to the ohmic resistance of the primary coil.

The co-efficient of mutual induction of two coils is equal to the number of lines of force passing through the secondary winding multiplied by the number of turns in that winding, when there is a unit current flowing through the primary winding. The practical unit of mutual induction and self-induction is the henry. The co-efficient of mutual induction in henries multiplied by the rate of change of the primary current in amperes per second gives the electromotive force induced in the secondary winding in volts.

From the above it will be seen that by winding a magnetic core with one coil of a small number of turns and another of a large number of turns, we may increase the voltage of an alternating current by making the former the primary, and decrease the voltage by reversing the functions of the two windings. The coil with a large number of turns is made of a comparatively fine wire, and the other with a comparatively heavy wire. It is obvious that the electrical energy appearing in the secondary circuit must be slightly less than that in the primary circuit, and, therefore, if the voltage of the primary is multiplied, say, ten times in the secondary, then the secondary current will at most be not quite one-tenth the primary current. Therefore, a much finer wire will do for the secondary winding. An apparatus of this type, comprising a magnetic core and primary and secondary windings, is known as an alternating current transformer (Fig. 46).

CHAPTER VI

Measuring Instruments

Movable Coil Type of Instrument.—The most commonly used type of electric measuring instrument is that known as the D' Arsonval or movable coil type. It consists of a permanent magnet of the horseshoe type, between the poles of which is arranged a coil of wire carried on a spindle which is delicately supported in jeweled or hardened bearings. Secured to the coil is an indicating hand adapted to move over a scale on a dial plate. The movable coil, whose ends are electrically connected to contact pins or binding posts on the outside of the instrument, is held in the zero position by a spiral spring. When a current is sent through the coil it makes the latter a magnet whose poles react with those of the permanent magnet. The principle of the instrument is illustrated in Fig. 47. If we suppose that the current flows through the coil in such a direction as to make the top end of the coil the north pole, as indicated, then, since like poles repel and unlike poles attract each other, the coil tends to turn around in a clockwise direction, as indicated by the arrow. If there was no resistance to the turning motion of the coil it would continue until its axis was in line with the magnetic lines of force of the permanent magnet. In an actual instrument (Fig. 48), to assume this position it must overcome the spiral spring, the force of which increases with the angle through which it is wound up. The force tending to turn the coil around depends, of course, upon the current flowing through it; hence the greater the current the greater the turning motion of the coil and of the indicating hand secured to it. Thus, by properly graduating the scale, the instrument can be made to indicate the current flowing through it. Such an instrument, whose dial is graduated in amperes, is known as an ammeter. However, the full current flows through the coil of the ammeter only in case the currents which the instrument is intendent to measure are comparatively small. If heavy currents are to be measured use is made of the principle that when a current may flow through

MEASURING INSTRUMENTS. 63

two paths in parallel it divides between them in the inverse proportion of their resistances. For instance, referring to Fig. 49, suppose that the coil of the instrument has a re-

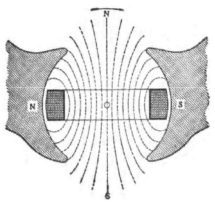

FIG. 47—ILLUSTRATING PRINCIPLE OF MOVABLE COIL INSTRUMENT.

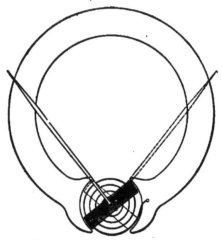

FIG. 48—DIAGRAM OF MOVABLE COIL INSTRUMENT.

sistance of 99 ohms and is shunted by a wire having a resistance of 1 ohm. Then the current passing through the instrument coil will be 1/99 of that passing through the shunt and 1/100 of the total current passing through the two together. As the deflection of the indicator hand depends upon the current passing through the coil, by placing the shunt across the coil a current 100 times as strong will have to be sent through the circuit to produce the same deflection as before. Any other multiplying factor may be obtained by using a shunt of the proper resistance. The current passing through the instrument coil is always a certain definite fraction of the total current in circuit and the scale can be graduated to show the total current.

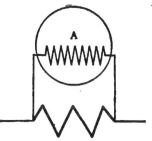

FIG. 49—ILLUSTRATING PRINCIPLE OF AMMETER SHUNT.

An instrument of exactly the same construction is used to measure electromotive force. In this instrument, graduated in volts and known as a voltmeter, the coil is wound with extremely fine wire and is connected between the points whose difference of potential is to be measured. The deflections, as before, are dependent upon the current flowing through the coil, but inasmuch as by Ohm's law

E. M. F. = Current $\times$ Resistance,

and the resistance of the coil is constant, the indications are proportional to the electromotive force applied to the ends of the coil.

Solenoid Type of Meter—The principle of this type of instrument is illustrated in Fig. 50. The current flows through a solenoid which is curved to form an arc of about 120 degrees just inside the housing of the instrument. Into this solenoid extends a piece of soft iron forming substantially a semicircle. One end of this soft iron core is secured to the spindle of the instrument which also carries the indicating hand. The soft iron core is nor-

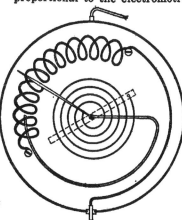

FIG. 50—SOLENOID TYPE OF INSTRUMENT

mally held in the zero position by a spiral spring. When in this position the free end of the core extends barely into the solenoid. It has been explained that a solenoid has a tendency to draw an iron core completely into it, or to such a position that the center of the core coincides with the center of the solenoid. The stronger the current flowing in the solenoid the greater the force tending to draw the core in and, consequently, the farther the spindle of the instrument will be turned around against the torsional resisting force of the spiral spring. Hence the motion of the pointer, which is secured to the spindle, is a measure of the current flowing through the instrument.

In some types of small instruments the scale divisions are not uniform throughout, but are smaller at one end than at

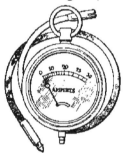 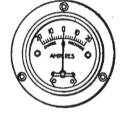

FIG. 51—POCKET TYPE BATTERY AMMETER.   FIG. 52—DASHBOARD TYPE AMMETER.

the other, or smaller at the ends than in the middle. It is desirable, of course, to have fairly uniform scale divisions, as readings of equal accuracy can then be made of any current or voltage within the range of the instrument. To get a uniform scale in the solenoid type of instrument the convolutions of the solenoid may be unevenly spaced along its length.

Pocket and Dashboard Instruments.—Electric measuring instruments used in connection with gasoline cars are of two kinds, viz., pocket instruments and dashboard instruments. Fig. 51 shows a pocket ammeter and Fig. 52 a dashboard voltmeter. Pocket instruments are generally provided with one spur which is insulated in the housing of the instrument and to which one end of the coil or solenoid is connected, and one short length of insulated flexible cable with a contact pin at the end. By means of the spur and contact pin the meter can be quickly electrically connected to a battery or cell to be tested. Dashboard instruments are provided with two binding posts to which the connections can be made. Some types of instruments, like the movable coil type, require that connections be made in a certaain way, in order that the instrument may give indications, while in other types, like the solenoid type, it is immaterial in which direction the current flows through. In the former types of instrument the two terminals are marked with the + and — sign respectively.

Connections of Instruments.—Electric circuits in general consist of a source of current such as a battery, one or more current consuming devices, such as lamps, and wires connecting the source of current and the consuming devices. Electrical instruments are used to determine the current flowing in the circuit and the electromotive force impressed on the circuit by the source. In Fig. 53 are illustrated the methods of connecting an ammeter and a voltmeter, respectively, to determine these factors. To measure the current the circuit is broken at some place and the ammeter (A) is inserted. To measure the voltage the voltmeter (V) is connected across the circuit, as shown. Of course, the voltage of a cell or battery can also be measured when no current consuming devices are connected to it and the contact points of the voltmeter are then pressed in contact with the two terminals of the cell or battery.

In testing dry cells a pocket ammeter is used whose contact pins are pressed in contact with the terminals of the cell exactly as would the contact pins of a voltmeter. The indication thus obtained, of course, will not show the amount of current which the cell will deliver to any particular consuming device, but the current it will give on "short circuit," or which it will force through its own internal resistance, because in comparison with the latter the resistance of the ammeter is practically nil, and the cell may be considered closed upon itself.

Short Circuits and Grounds.—Since the term "short circuit" is here used, it may be well to explain its meaning as applied in electrical work. If in a circuit as depicted in Fig. 53 the two wires leading from the battery to the consuming devices were electrically connected, at some point of their length, by a short, heavy wire, or were brought in direct contact by being placed side by side and the insulation chafed off, then there would be established a "short circuit" —one of the most common faults in electrical work. The effect of such a short circuit as just described would be that a very heavy current would be drawn from the battery,

MEASURING INSTRUMENTS.

which would flow merely through the wires, and practically no current would flow through the consuming devices, as the resistance of the consuming devices would be very great as compared with that of the alternate path through the short circuit. This is a true short circuit. However, the term "short circuit" is applied to any undesired or accidental leakage path, whether of high or low resistance.

Generally all the wires forming part of an electric circuit, as well as the devices connected by them, are insulated. Where electric conductors are supported on a metallic framework, as in the case of an automobile, it will occur that the insulation at some part of the circuit becomes damaged and the conductor comes into metallic contact with the framework. The circuit is then said to be "grounded" (the English use the term "earthed"). A ground is indicated diagrammatically as shown in Fig. 54. All of the parts of the

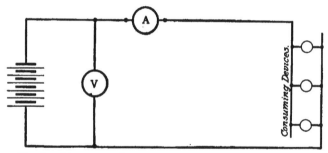

FIG. 53—METHOD OF CONNECTING VOLTMETER AND AMMETER.

automobile chassis are of good conducting material and of heavy section, consequently they offer very little resistance to the passage of current. Frequently the framework of the car is used for one side of a circuit. That is to say, the current flows from the source through an insulated wire to the consuming device. After passing through the consuming device it passes through a "ground connection" into the frame of the car and through another ground connection it returns to the source. An electric wiring system employing the car frame for the return conductor is known as a "single wire" or "ground return" system. Portions of a circuit often become grounded accidentally. A single ground on a "single wire" and grounds on opposite sides of a two wire system are equivalent to a short circuit.

Use of Instruments.—In using electrical instruments, care should be taken not to expose them to currents or electro-

68   MEASURING INSTRUMENTS.

FIG. 54—DIAGRAM OF GROUND CONNECTION.

motive forces far beyond their range, as this would be injurious. An ammeter, for instance, should never be connected directly to the terminals of a storage cell, as a storage cell thus short circuited passes a current far beyond the capacity of any of the usual automobile instruments. For testing storage cells as to state of charge, etc., a voltmeter should be used, in connection with the voltage curves (Fig. 15). Voltage readings of a storage battery should be taken only when the battery is charging or discharging, as readings taken when the battery is idle are apt to be misleading. For testing dry cells the ammeter is the best instrument, but in making the test it should be applied to the cell only long enough for the pointer to come to rest, as the heavy drain on the cell is injurious to it. Ammeters for this purpose usually have a range up to 30 amperes. Voltmeters for ignition batteries have a scale up to 10 volts. However, voltmeters are made with many different ranges, and often they have two scales arranged concentrically. For instance, suppose that the upper or outer scale has a range 0-3 volts. If now we insert in the instrument a resistance equal to twice the resistance of the instrument coil and connect it in series with the latter, then, of course, the total resistance of the instrument being trebled, it requires three times the electromotive force as before to send a certain current through the instrument coil and cause a certain deflection of the needle. Therefore, the second scale will be three times as high as the first and will have a range of 9 volts. This would make the instrument suitable for accurately testing both single cells and 6-7 volt batteries. Of course, the instrument would be provided with a third binding post between which and the second binding post the resistance is connected.

Double Range Instruments—Voltammeter.—Ammeters are also made with double scales. For one scale all of the current measured may pass through the coil or solenoid and for the other the coil or solenoid may be shunted, as already explained. In making connections to a dashboard ammeter, wires of a cross section easily able to carry the maximum current must be used.

Ammeters and voltmeters are also combined into a single instrument and are then known as voltammeters. Sometimes these combined instruments possess two separate movements, but more frequently they have only separate coils, the same movement showing both volts and amperes. In the latter case there are three binding posts of which one is used for both volt and ampere readings and the other two are used for volt and ampere readings respectively.

CHAPTER VII

The Magneto Generator

In the early years of electric ignition, when the motors usually had only one or two cylinders, batteries served almost exclusively as sources of ignition current. But as the number of cylinders was increased to four, and higher engine speeds also became customary, the drain on the batteries became so great that they had to be renewed or recharged very frequently. This led to the adoption of mechanical generators, chiefly magnetos.

Construction.—A magneto generator consists essentially of two parts, the magnet frame and the armature. The former is the stationary, the latter the rotary part. The magnet frame (Fig. 55) is built up of a number of U-shaped magnets (though they are generally referred to as horseshoe magnets), from two upward, the individual magnets being placed either all side by side or made in two sets, of which one straddles the other. U-shaped magnets of rectangular cross section are found in most magnetos and may be regarded as standard, but there are

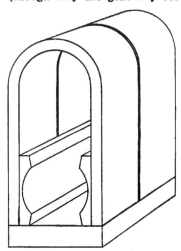

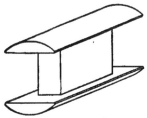

FIG. 55 — PERMANENT MAGNETS, POLE SHOES AND BASE.

FIG. 56—SKETCH SHOWING SHAPE OF ARMATURE CORE.

a few makes of magnetos that have magnets of different shape. To the ends or poles of the magnets are secured pole pieces of soft cast iron or drawn steel, which are bored out to form an armature tunnel. At the bottom of the magnets is a base plate of nonmagnetic material (brass). The armature core is of the shuttle or H type, as shown in Fig. 56. Its central portion is built up of soft sheet iron discs which are held between end plates with lateral projections. The discs and end plates are held

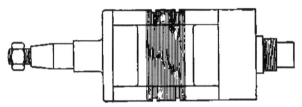

FIG. 57—ARMATURE CORE WITH END PLATES AND SHAFT.

together by rivets. This soft iron core serves as a bridge for the magnetic lines of force in passing from one pole shoe to the other and also to carry the winding in which the current is induced. Before assembling, the discs are varnished so they will be practically insulated from each other. After the core is assembled it is thoroughly insulated and the wire is wound on. In practically all ignition magnetos one end of the armature winding is electrically connected to the armature core (grounded), while the other end is brought out to a collecting device.

After the winding of the armature has been completed, two bands of steel wire are put around the armature, in grooves provided therefor in the core (see Fig. 57). These bands are tinned and clipped, and serve the purpose of preventing the armature wire from flying out and coming in contact with the magnet poles, under the action of centrifugal force when the armature is rotating at high speed.

Principle of Action.—The action of the magneto generator will be readily understood from the principles explained under Electro Magnetic Induction. Normally, when at rest and free to turn, the armature rests in what may be called a horizontal position (A, Fig. 58). The path for the magnetic lines of force is then the easiest, and the magnetic flux through the armature is a maximum. For this reason the core has a strong tendency to return to this position when moved away from it. If now the armature be turned around its axis there will at first be little change in the number of lines of force through it, for as long as a considerable portion of the armature poles is opposite the field poles there is little change in the magnetic reluctance of the circuit. But as the armature poles move out from under the field poles (B, Fig. 58), the reluctance of the magnetic circuit increases and consequently the flux decreases. When the armature core is in a vertical position no lines of force pass through it from one of its poles to the other. This is the point where the direction of the flux through the core reverses. Up to this point the lines have entered the core through what is now the upper pole piece of the armature and left through the lower, but when the core has passed

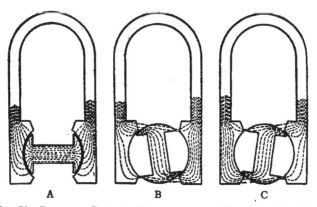

Fig. 58—Diagrams Showing Distribution of Magnetic Flux for Various Armature Positions.

slightly farther on (C, Fig. 58) the lines enter through the bottom pole piece and leave through the top one. The induced electromotive force at any instant is proportional to the rate of change in the number of lines, and is a maximum (if the armature circuit is open, so that no current is flowing) when the armature core is in the vertical position. When the armature core is passing through the horizontal position (A, Fig. 58) there is no change in the number of lines of force through it and the induced electromotive force is nil. Since the armature passes twice through the horizontal position during each revolution, the induced e.m.f. is nil twice during each revolution, and it also passes through two maxima, a positive and a negative one. Fig. 59 shows the curve of the e.m.f. induced in a magneto armature when

THE MAGNETO GENERATOR.

no current is flowing. It is obvious that a magneto of this type gives an alternating current.

Armature Reaction.—When the magneto is in service there is always a current flowing in the armature and this current has a certain influence on the magnetic flux through the armature core. The armature current sets up a field in a vertical direction and this combines with the field due to the permanent magnets as shown in the diagram Fig. 60. The flux due to the permanent magnets is a maximum when the armature core is in position A and the flux due to the armature current is a maximum 90 degrees later when the armature core is in a vertical position. In other words, the latter lags 90 degrees behind the former. In Fig. 61 the

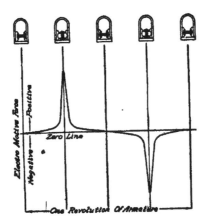

FIG. 59—CURVE OF PRIMARY E. M. F. ON OPEN CIRCUIT.

horizontal line represents the flux due to the permanent magnets and the vertical line the flux due to the armature current. The inclined line represents the resultant of the two both in direction and magnitude. We see from this figure that the effect of the armature current is to shift the magnetic field around in the direction of armature rotation, and, in consequence, the induced e.m.f. attains its maximum value not while the armature core is in a vertical position but a moment later.

The armature is usually carried in ball bearings mounted in end plates secured to the magnet frame and base. The end of the armature winding is brought out to a collector pin passing through the armature shaft but insulated from it, on which bears a carbon brush carried in an insulated brush holder and pressed against the pin by a flat spring. The brush holder connects with a binding post to which a wire or cable connection can be made.

Variations in Size.—Magnetos vary a good deal in size and weight, and as a rule the smallest magnetos are found on the smallest motors. At first sight it seems strange that a larger magneto should be required to fire a large motor, but on further consideration there is found to be good reason for the practice referred to. To insure positive ignition the magneto must generate a certain minimum electromotive force. The e.m.f. which a magneto generates is proportional to the magnetic flux through its armature, to the number of turns on the armature and to its speed of rotation. In a large magneto the flux through the armature and the number of turns will be greater, hence the minimum speed at which the magneto will produce an effective spark will be less. But a big motor of many cylinders cannot be cranked as fast as a small one and it generally also is required to have a lower idling speed, hence a magneto for such a motor must be capable of generating a good spark at very low speed, and, in consequence, must be comparatively large.

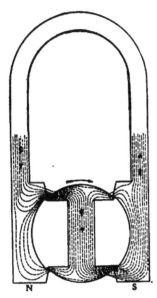

Fig. 60—Diagram Showing Armature Cross Magnetization.

Some of the dimensions of magnetos have been standardized by the Society of Automotive Engineers with the object of making magnetos of different makes interchangeable on the same base.

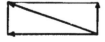

Fig. 61—Combining Magnetic Effects Due to Permanent Magnets and Armature Current.

THE MAGNETO GENERATOR.

S. A. E. MAGNETO STANDARDS.

	Mm.	Ins.
Shaft height	45	1.771
Distance from center of front base plate holes to large end of shaft taper	53	2.086
Distance from center of front to center of rear base plate holes	50	1.968
Distance between centers of base plate holes transversely	50	1.968
Large diameter of taper	15	0.590
Small diameter of taper	12	0.472
Length of taper	15	0.590

Taper 1:5—11° 30′ approximately.
Woodruff key No. 3.

Height of magneto space	8.00
Length of magneto space	10.00
Width of magneto space	5.00
Plain hole of timer lever	**0.25**
Tapped hole of timer lever	¼—28
Base plate holes	⅜—16 U. S. S.
Thread at end of shaft	⅜—16
Length of thread	0.5905″
Length of advance lever (center to center)	2⅛ inches.

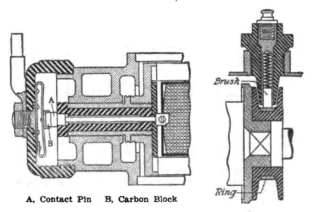

A, Contact Pin   B, Carbon Block

FIG. 62—TWO FORMS OF MAGNETO CURRENT COLLECTORS.

CHAPTER VIII

Low Tension Ignition

So far as automobile practice is concerned, low tension ignition is a thing of the past, but it may prove of interest from an historical standpoint to briefly review the low tension systems formerly used for automobile ignition. The simplest was the battery low tension system used for single or double cylinder motors.

Low Tension Spark Plug.—With such a system there is used a pair of electric contact terminals passing through a flanged block in the cylinder wall. Fig. 63 shows such a block, known as a low tension spark plug. It carries an insulated stationary terminal and a non-insulated movable terminal, both of which are sometimes provided with inserted contact points of some metal that does not readily oxidize at high temperatures. By a mechanism operated from the camshaft the two terminals are allowed to be brought, by means of a spring, into contact inside the cylinder an instant before the spark is to take place, and are then abruptly drawn apart. The source of current may be either a battery of 6-10 volts or a magneto generator. Included in the circuit with the battery and the spark plug is a choke coil or low tension spark coil. This consists merely of a core of soft iron wire about 1 inch in diameter and 10 inches long, over which is passed an insulated spool which is wound with

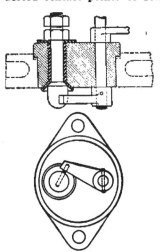

FIG. 63—LOW TENSION SPARK PLUG.

LOW TENSION IGNITION.

insulated copper wire, say three layers of about No. 14 B. & S. for a voltage of 8. The electrical circuit is sketched in Fig. 64, which also shows the switch for closing and opening it.

It will be understood that when the switch is closed and the engine is then turned until the spark terminals come in contact, a current will flow through the circuit. When the latter has attained nearly its full value the circuit is sud-

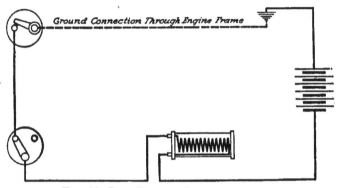

FIG. 64—LOW TENSION IGNITION CIRCUIT.

denly interrupted by the terminals breaking contact, whereupon a powerful spark is produced at the point of the break by the extra current resulting from the discharge of the energy stored up in the coil.

Make-and-Break Mechanism.—Fig 65 shows one form of mechanism used for actuating the spark terminals. To the outer end of the movable terminal is secured a short lever arm A whose end passes over a vertical tappet rod B. The latter is forced upward by a cam C, thereby bringing the spark terminals inside the cylinder in contact. The excess motion of the tappet is taken up by the coiled spring D underneath the terminal arm. When the cam passes by the foot of the tappet B the latter is quickly lowered, owing to the pressure of the coiled spring E and its own weight, and the contact points are then separated.

Fig. 66 shows a low tension ignition mechanism for a four-cylinder engine. The make-and-break mechanism here illustrated is slightly different from that just described. Cam G forces rod F upward against the pressure of spring H and allows the lighter spring E to draw the spark terminals together. Then the cam trips the rod and, owing to the pres-

LOW TENSION IGNITION.

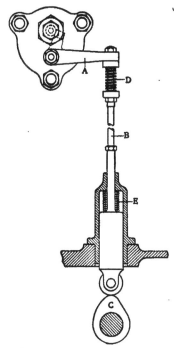

FIG. 65 — MAKE-AND-BREAK MECHANISM.

sure of spring H, a blow is dealt the arm of the movable terminal, which causes it to break contact sharply.

Spark Timing—The moment at which the upward pressure on the rod is released and the downward pressure begins is definitely fixed and always corresponds to a certain position of the engine crankshaft. However, the spark occurs at a somewhat later period, when the two spark terminals begin to separate. The time interval between the moment when the peak of the cam passes from underneath the igniter rod and the moment when the terminals begin to separate depends upon the force of spring H and the inertia of the moving parts of the igniter mechanism, and is constant. But this constant time interval or igniter lag corresponds to a greater angular motion of the crankshaft the higher the speed of the engine. Therefore, if the igniter rod were tripped at exactly the same moment, whatever the speed of the engine, the spark would occur later in the cycle of the engine when the latter was running at high speed than when it was running at low speed, which is the direct opposite of what is desirable. To make the spark always occur at exactly the same period in the cycle of the engine or to advance it slightly as the engine speed increases it is necessary to trip the igniter rod earlier when the engine runs at high speed. This is accomplished by means of the bar L controlled by a bell crank by means of which the lower end of the igniter rod may be moved toward the right so that the igniter cam will trip it earlier in its revolution. This spark advancing mechanism is controlled by means of a lever at the driver's seat so that the driver can advance and retard the time of the spark at will as the engine is running.

LOW TENSION IGNITION.

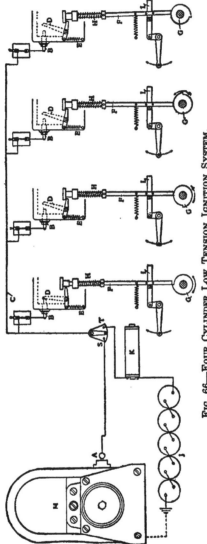

FIG. 66—FOUR CYLINDER LOW TENSION IGNITION SYSTEM.

It will be observed that all of the spark terminals are normally apart. Each pair makes contact only a moment before the spark is to be produced, and it is only during this short interval that the circuit is closed and the arc between contact points is maintained, that current is flowing.

In Fig. 66 are shown two sources of current, the magneto M and the battery J, either of which can be used at will by means of the double throw switch ST. The spark coil K is used only in connection with the battery, as the magneto armature has sufficient self-induction to produce a powerful spark when its circuit is opened.

The chief reason why the low tension ignition system was discarded was undoubtedly that it involved the use of considerable delicate mechanism which was prone to be noisy. The complication of the mechanism in the case of a multi-cylinder engine may be judged from Fig. 66. Contact points, of course, would wear out and sometimes would become short circuited by carbon and oil inside the cylinder.

Magnetic Spark Plugs.—In another system of low tension ignition magnetically operated spark plugs take the place of the mechanically operated plugs. A well-known magnetic spark plug is the Bosch, of which a sectional view is shown herewith (Fig. 67). This consists primarily of three parts, a so-called iron clad electro magnet, a support which screws into the spark plug aperture in the engine cylinder, and the oscillating portions. The electro magnet consists of a shell A with cover B containing a coil D. Through this coil projects a cylinder H, which is threaded at its lower end. A collar R with round neck forms the base of the magnet.

The oscillating portions consist of three parts—a pole piece E, an armature F and a horseshoe shaped spring G. The lower part of E is threaded to fit the hollow cylinder H and is exteriorly formed to fit inside the support K, in which it is held by a screw collar L. The support is made with the upper half hexagonal and the lower half threaded to fit the spark plug aper-

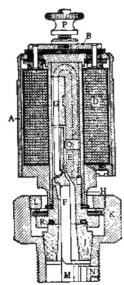

FIG. 67—SECTIONAL VIEW OF MAGNETIC SPARK PLUG.

ture. A steatite plug J in connection with gaskets insures gas-tightness.

The terminal P having been connected to the source of current, when the current passes the coil D magnetizes core E. This draws armature F (which is pivoted on a knife edge on E) to the right, breaking the circuit between M and N, causing a spark to pass. A brass piece O is placed so that the armature abutts against it, thus preventing "sticking" between armature and pole piece due to residual magnetism. Spring G tends to keep points M and N in contact. N is V-shaped, and M on the armature is ground to the same angle as the V portion of N. The source of current is a low tension magneto.

Magnetic Plug System.—With the magnetic spark plug, the same as with the make-and-break mechanism, there is a certain time lag which causes the spark to occur later in the cycle when the engine is running at high speed. The fixed point is that at which the electric circuit is closed, which occurs always at the same point in the cycle of the engine. The current in the circuit must then grow to such a value that the attraction of the magnet core overcomes the force of the spring and the inertia of the movable part of the plug, which takes a certain definite time. In order to insure that ignition may occur always at the same point in the cycle of the engine, the device which closes the circuit must be so arranged that the time of closing can be advanced as the speed of the engine increases.

In addition to the timer, the circuit of a magnetic spark

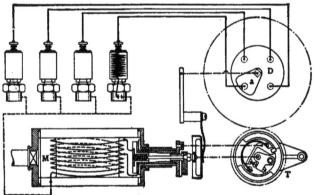

Fig. 68—Diagram of Bosch Magnetic Spark Plug Ignition System.

plug ignition system for multi-cylinder engines must contain a distributer which will cut the different magnetic spark plugs in and out of circuit with the source of current in the order in which they must spark or fire the charge. A diagram of the complete ignition system employing magnetic spark plugs, for a four-cylinder engine, is shown in Fig. 68. In this cut M is the low tension magneto and T the interrupter or timer by which the circuit can be closed at an earlier or later point (within a certain range) in the engine cycle. This forms an integral part of the magneto. An almost identical device is used in connection with modern high tension magnetos and a detailed description of the device will be given later on.

The distributer, which makes connection successively to the different spark plugs, is designated by D. It consists of four contact points carried by a block of insulating material, and a rotating arm A which makes contact with the four points in turn.

CHAPTER IX

Elements of a High Tension Battery System

There are two distinct systems of electric ignition for internal combustion engines, viz., the low tension system already described, in which the arc which always accompanies the breaking of a current-carrying circuit is made use of, and the high tension system in which a sudden impulse of induced electromotive force causes a spark between terminals which come close together inside the engine cylinder. In low tension ignition the electromotive forces in play are comparatively small. The e.m.f. of the source is generally only about six volts, and though the e.m.f. of self-induction, which causes the spark, considerably exceeds this, it is far lower in value than the e.m.f. required to cause a spark between two terminals at a distance of 1/32 inch. In the low tension circuit, as the circuit is broken, the current volatilizes a small portion of the metal of the positive terminal, and this metal in the gasified state forms a conducting path of relatively low resistance; hence a comparatively low e.m.f. suffices to maintain the arc.

The Induction Coil—To obtain the high electromotive force necessary with the high tension system to break down the resistance of the air gap, use is made of a form of transformer known as a high tension coil or induction coil. It consists of a cylindrical core of soft iron wire over which is wound a primary winding of relatively coarse wire and a secondary winding of very fine wire, the latter having perhaps a hundred times the number of turns as the former. The primary winding of this coil is connected in circuit with the source of current, an interrupter and a switch. (See Fig. 69.) The secondary is connected in circuit with a spark plug which contains two terminals which are insulated from each other. The spark plug is screwed into the wall of the combustion chamber and its spark terminals come within a distance of 1/32 inch or less of each other inside the chamber.

The Mechanical Vibrator—The interrupter is actuated from the engine crankshaft, so that it operates in synchronism

84 ELEMENTS OF A HIGH TENSION BATTERY SYSTEM.

with the latter. The illustration, Fig. 69, shows a form of interrupter called a trembler or mechanical vibrator, which was much used in early years. It consists of a cam A with a triangular slot at one part of its circumference, a flat spring B with a nose at its free end, and an insulated, adjustable contact screw C, the point of which is located opposite a contact piece on spring B. Cam A is rotated at one-half the speed of the engine crankshaft, through so-called 2:1 gearing, a pinion on the crankshaft engaging with a gear of twice its number of teeth on the cam shaft. When the nose of the spring rides on the cylindrical surface of the cam the two contact points are out of contact, but when the slot on the cam comes opposite the nose on the spring, the flexure of the latter brings the

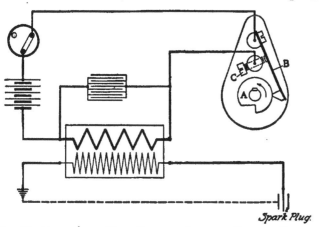

FIG. 69—DIAGRAM OF HIGH TENSION COIL AND BATTERY SYSTEM.

points in contact. In fact, the spring is set in vibration and contact is made and broken a number of times in quick succession.

"Make" and "Break" Induced E. M. Fs.—Upon the circuit being closed by the interrupter a current will be forced through the primary winding of the coil by the battery, this current increasing gradually, as illustrated by the curve shown in Fig. 39. The increase of the primary current induces an electromotive force in the secondary winding which is opposite in direction to the primary current. When the primary circuit is broken by the interrupter the primary current decreases, and this decrease results in another impulse of electromotive force in the secondary winding, this one being in the

ELEMENTS OF A HIGH TENSION BATTERY SYSTEM. 85

same direction as the primary current. The magnitude of this e.m.f. depends upon the number of turns in the secondary winding and also upon the rate of decrease of the primary current. The self-induction of the primary circuit opposes a quick change in the magnitude of the current and upon the breaking of the circuit causes a spark to appear at the contact points of the interrupter. To prevent this spark and insure a quicker cessation of the primary current a condenser is connected in shunt to the interrupter to absorb the extra current of the primary circuit. This is indicated in Fig. 69.

Owing to the provision of the condenser, the primary current decreases much faster upon the circuit being opened than it increases upon the circuit being closed, and, consequently, the secondary induced electromotive force corresponding to the break is much greater than that corresponding to the make of the primary circuit. The way high tension ignition systems are generally designed and operated, the secondary e.m.f. corresponding to the closing of the primary circuit is insufficient to cause a spark to jump the gap between the terminals of the spark plug, and it is the e.m.f. corresponding to the *opening* of the primary circuit that is utilized for sparking.

To produce a disruptive discharge between terminals separated by an air space requires an enormous e.m.f., which varies with the length of the air gap and also depends to a certain extent upon the form of the terminals. It is of the order of 3000 volts per 1/16 inch. The following table shows the voltages required to produce a spark between metal balls 1 centimeter in diameter, when different distances apart in the atmosphere:

Distances Apart (cm.)	Volts Required
0.1	4.830
0.5	16.890
1.0	25.440
1.5	29.340
2.0	31.350
3.0	37.200
5.0	45.900
10.0	56.100
15.0	61.800

In ignition work a spark has to be forced through a gap filled with a gaseous mixture under 5 to 7 atmospheres compression, and the compression greatly increases the resistance to the spark, hence calls for a still higher e.m.f. The relation between the compression pressure and the e.m.f. required to force a spark across a certain gap is shown in Fig. 70.

The form of interrupter shown in Fig. 69 was soon aban-

86 ELEMENTS OF A HIGH TENSION BATTERY SYSTEM.

doned, on account of the difficulty of keeping it adjusted. Its place was taken by a so-called magnetic vibrator or buzzer which is used in connection with a timer. The object of the latter is merely to close the circuit for a certain period during each cycle of the engine. As soon as the circuit is closed by the timer the magnetic vibrator begins to operate, opening and closing the circuit a number of times and producing a series of sparks in the cylinder.

Magnetic Vibrator—Fig. 71 shows a typical design of magnetic vibrator. It consists of a flat steel spring A, which has riveted to it a disc B of soft iron, which acts as armature to the core C of the induction coil. The flat spring is rigidly

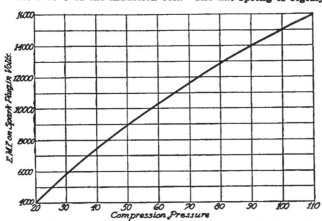

FIG. 70—VARIATION OF SPARK VOLTAGE ACROSS A 0.020 INCH GAP WITH COMPRESSION.

held at one end by the support D mounted on the end of the coil, and its other end carries the contact point c, which is normally in contact with a similar point c¹, carried by a bridge E, also mounted on the end of the coil box. Between support D and armature B there is an adjusting nut F, which passes through a slot in spring A and is adapted to move up and down on a screw secured in the head of the coil box. The spring rests on a shoulder of this nut and is thereby placed under tension so as to cause the contact points c and c¹ to make contact. By unscrewing nut F the pressure at the contact points can be increased, and vice versa, and the adjusting nut F is automatically locked in any position by means of a spring steel latch secured to the support D, engaging into notches on a flange nut F.

ELEMENTS OF A HIGH TENSION BATTERY SYSTEM. 87

Normally the two contact points are in contact, so that there is no break in the circuit at this point. If now the primary circuit is closed by the timer, a current will start to flow through the primary winding of the coil, whereby the core of the coil will become magnetized. It will then attract armature B against the force of spring A, whereby the circuit is broken at the contact points c. c^1. As soon as the circuit is broken the current ceases to flow through the primary winding of the coil, the core loses its magnetism and armature B recedes from the end of the core under the tension of the spring. Contact is then re-established at points c c^1, the current flows once more through the primary winding and attracts the armature, there-

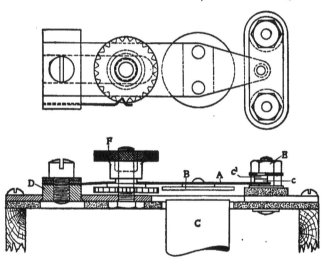

FIG. 71—MAGNETIC VIBRATOR.

by again breaking the circuit. This continues as long as the circuit remains closed at the timer. These magnetic vibrators have a periodicity of from 100 to 400 per second, depending upon dimensions and other features, 200 per second being a good average.

Construction of Coil—The central part of the spark coil is the core, which is made of very thin, highly annealed iron wire. It is necessary to anneal the iron because it has to be magnetized and demagnetized at a very rapid rate, and if it were not made of very soft iron there would be considerable hysteresis loss. Ignition coils vary considerably in size, but in the two most common sizes the core is 5 and 6 inches long and ⅝ and ¾ inch in diameter, respectively, the usual ratio of length to diameter being 8. Upon the core is wound the primary winding, which consists of so-called magnet wire, usually double cotton covered, two layers of No. 16-18 B. & S. gauge. The primary winding is then covered with a layer of insulating material, which in some cases is in the form of a hard rubber tube, and over this is placed the secondary winding, which consists of some 20,000 turns of very fine wire, about No. 36, silk covered. The writer has on hand data of a great many coils wound for operation on 4 and 6 dry cells, in which the primary resistance ranged between ¼ and ⅜ ohm and the secondary resistance between 2000 and 4000 ohms. While the secondary winding may have only about one hundred times as many turns as the primary, it must not be inferred from statements made under the heading of Electromagnetic Induction that the secondary induced electromotive force will be

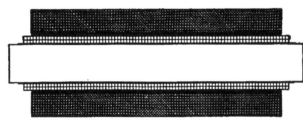

FIG. 72—SECTIONAL VIEW OF COIL.

only one hundred times that impressed upon the primary winding. This holds good only in case an alternating current flows through the primary winding. When a primary current is suddenly interrupted, as in this case, the ratio of the maximum value of the secondary induced e.m.f. to the e.m.f. which caused the current in the primary circuit is far greater than the ratio of the number of turns.

The secondary is usually wound separately and after suitable treatment is placed in position over the primary. Great care has to be used to insure perfect insulation of the secondary winding, because the electromotive force between its terminals when in operation is sufficient to cause a spark in the atmosphere about ½ inch long.

Vacuum Impregnating Process—To further improve the insulation of the coil it is impregnated with paraffin or with a composition of beeswax and rosin, about one part of the former to three of the latter. Formerly this impregnating process was applied in the atmosphere, but the coils always held some moisture in the insulating material and air bubbles or traps would often form inside the coil, excluding the insulating composition and thus forming weak spots. To obviate this, coils are now impregnated in vacuum. In applying the vacuum impregnating process, use is made of a vessel of cast iron or heavy sheet steel with a cover screwed on to make it air-tight. Inside the vessel is a coil of steam pipe and live steam is sent through this coil to heat the vessel. After the coils have been placed inside and the vessel has been closed, the steam is turned on and at the same time the air is exhausted from the vessel by means of an air pump connected with it. Between the air pump and the vessel is arranged a condenser in which any moisture held by the air in the vessel condenses. The coils are subjected to the heat in the vacuum for a couple of hours, which insures the removal of all air and moisture from them. Then the insulating compound, which has been melted in another adjacent vessel, is drawn into the impregnating vessel through a valve, without breaking the vacuum in the latter. After the coils have been immersed in the insulating compound for a sufficient length of time to make sure that the latter has well penetrated, the air pump is run so as to act as a compressor and force air into the impregnating vessel until a pressure of four atmospheres or so is reached. This makes complete penetration of the insulating compound doubly sure. Then the valve through which the insulating compound entered the impregnating chamber is opened and the compound is forced out by the air pressure.

The Condenser—In order to insure a quick cessation of the primary current and prevent destructive arcing at the vibrator contact points a condenser is always incorporated with an induction coil. A condenser is made up of sheets of tin foil (conducting material) and sheets of paraffined paper, or, where compactness is very essential, mica. In making a condenser, rectangular sheets of paraffined paper and of tin foil are first prepared in considerable number, all sheets of either material being alike. The paper is cut to such a size that it will overlap the tin foil all around, except at one side or one corner. The individual sheets are then piled on top of one another as follows: A sheet of paper, a sheet of tin foil overlapping the paper at one side or corner, a sheet of paper, a sheet of tin foil overlapping on the opposite side or corner, a sheet of paper, a sheet of tin foil overlapping on the same side as the first, and so on—that is, always a sheet of paper between successive sheets of tin foil and successive sheets of tin foil overlapping the paper on opposite sides. This is illustrated

90 ELEMENTS OF A HIGH TENSION BATTERY SYSTEM.

in Fig. 73. After all of the sheets have been assembled they are bound together by means of rubber bands or tape, the overlapping portions of the tin foil are riveted together and leads of stranded wire are connected to the rivets. It may here be pointed out that the leads from the primary and secondary coil are also made of stranded wire so there is no danger of breaking them off.

Of late condensers are generally being wound of strips of tin foil and paper, taken from spools, this being a better production method than that described above. Two spools of tin foil and two of paper are used.

The coil, together with the condenser, is placed in a wooden box with dove-tailed joints, the condenser being either wrapped around the coil or placed beside it, in which latter case the

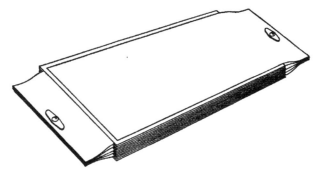

FIG. 73—METHOD OF BUILDING UP CONDENSER.

box has a rectangular cross section. The coil is anchored in the box by means of insulating compound poured into the box after the coil is in place.

Of course, the capacity of the condenser, which depends upon the aggregate surface of the tin-foil sheets directly opposed, upon the "closeness" of adjacent sheets and upon the character of the dielectric (insulating material) is proportioned to the dimensions of the coil itself. Its best value is determined in the first place experimentally by varying it until no spark is observable at the vibrator.

Connections of Coil—So far as the operation of the coil is concerned, there is no need for any electrical connection between the primary and secondary winding, the electrical energy being transferred from one to the other through the intermediary of the magnetic lines of force, which interlink both.

ELEMENTS OF A HIGH TENSION BATTERY SYSTEM. 91

However, for the sake of simplicity of external connections the beginning of the secondary winding is usually connected to the end of the primary winding. Both the primary and the secondary circuits have a "ground return" which necessitates that one end of both the primary and the secondary winding of the coil be placed in positive metallic connection with the engine or car frame. By connecting the primary and secondary windings as described, a single wire from the coil to the engine frame will serve to ground both windings. The average coil, therefore, has only three terminals, viz., one primary terminal, one secondary terminal and one common terminal.

The Timer—The function of the timer, as already explained, is to close the primary circuit during a certain portion

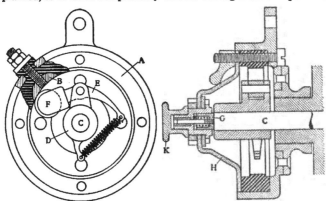

FIG. 74—ROLLER CONTACT TIMER.

of each cycle of the engine (two revolutions of a four stroke, one revolution of a two stroke engine). It consists of two parts, one stationary, the other revolving. The most popular type is the roller contact timer, which is illustrated in Fig. 74. It consists of a cylindrical housing A, of insulating material, in which is embedded a metal contact segment B, flush with the inner cylindrical surface. This segment carries a binding post projecting from the outside of the housing. Centrally within the housing is the end of the shaft C, which is driven through gearing from the crankshaft in such a ratio that it makes one revolution for every cycle of the engine, i. e., for every revolution of a two-stroke engine and for every two revolutions of a four-stroke engine. To this shaft is secured a hub D with a radial arm, at the end of which is fulcrumed a

92 ELEMENTS OF A HIGH TENSION BATTERY SYSTEM.

lever E, one arm of which carries a roller F, which is pressed in contact with the inner surface of the housing by means of a coiled spring extending between the end of the other arm and a lug on the hub. All of the rotating parts of the timer are grounded, because the shaft forms part of the metallic mass of the engine. However, this ground is somewhat unreliable, because there is a film of oil between the shaft and its bearing and also between the gear teeth; hence it is customary to provide a special ground connection. Referring to Fig. 74, a brush G is carried in the center of the cover plate H and is pressed by means of a spring against the end of shaft C. A grounding wire is run from the brush holder K to some point

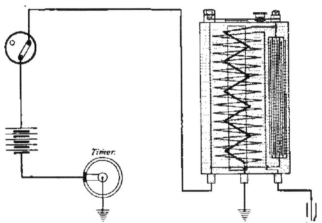

FIG. 75—DIAGRAM OF COIL AND BATTERY HIGH TENSION IGNITION SYSTEM WITH MAGNETIC VIBRATOR.

on the engine frame. When the roller is in contact with the metallic segment B, it completes the primary circuit.

Timer shafts are usually arranged vertically and extend as high as the top of the engine cylinders, where the timer is very accessible. The back or bottom of the timer housing is a casting and contains a bearing by means of which the timer is supported on the shaft. The top is a cover plate (usually aluminum) held in place by some quick acting locking device.

Primary and Secondary Circuits—Fig. 75 shows in diagram a complete high tension battery and coil ignition system for a single cylinder engine, in which a timer and magnetic vibrator are used. Both the primary and the secondary circuit are readily traced out in this diagram. Starting from the bat-

ELEMENTS OF A HIGH TENSION BATTERY SYSTEM. 93

tery the primary current flows through the switch, the magnetic vibrator, the primary winding of the coil, through the ground connection of the coil, through the engine frame, through the timer and back to the battery. The secondary current, originating in the secondary winding of the coil, flows through a high tension cable to the spark plug and thence through the engine frame and ground connection back to the secondary winding.

Why Variable Timing Is Needed—In the operation of this system the action of the ignition mechanism which is fixed in time relative to the cycle of the engine is the closing of the primary circuit by the timer. That is to say, whether the engine runs fast or slow, when the circuit is closed by the timer the engine crankshaft is always in a certain definite position. But between the moment the circuit is closed and the moment the spark passes there is a certain time interval. The current in the primary circuit rises gradually and it takes a certain time for it to attain such a value that it may overcome the force of the vibrator spring and open the circuit. This time interval between the closing of the circuit by the timer and the occurrence of the spark is constant, but it corresponds to a greater arc of revolution of the crankshaft when the engine is running at high speed than when it is running at low speed. It will thus be seen that the spark will occur later in the cycle when the engine is running fast than when it is running slowly. This is the exact opposite of what is required, because it takes a certain length of time for the flame, which starts from the spark terminals, to spread throughout the mass of combustible charge, which time is negligible in comparison with the time of a cycle at 500 revolutions per minute but not at 2000 revolutions per minute. Therefore, it is necessary that the moment of closing the primary circuit with respect to the cycle of the engine be advanced as the engine speed increases. This is accomplished by mounting the shell of the timer so it can be rocked around its axis. The housing is provided with a lug from which a lever and link connection is made to a "spark lever" on the steering hand wheel or on the steering post, so that the driver can vary the position of the timer housing from his seat. In Fig. 76 the crank arm is shown in full lines in the dead centre position, and the timer roller has a moment ago established contact with the timer segment, so that the spark may be assumed to occur at this moment, that is, when the crank is in the dead centre position. If now the timer housing be turned through a certain angle in the direction of the arrow, as indicated by the dotted outlines of the timer lug and terminal, then the spark in the

94 ELEMENTS OF A HIGH TENSION BATTERY SYSTEM.

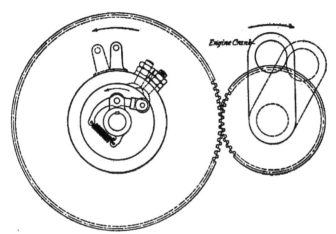

FIG. 76—ILLUSTRATING THE PRINCIPLE OF VARIABLE TIMING.

engine cylinder will occur only when the crank has assumed the position indicated by the dotted lines, that is considerably later. If the housing is turned in the direction in which the timer shaft revolves (as indicated by the arrow), the circuit is closed later in the cycle, and consequently the spark occurs later, and if the housing is revolved in the opposite direction the spark occurs earlier. The ignition lag, i.e., the time interval between the closing of the primary circuit and the beginning of the spark in the secondary circuit, amounts with different coils to from 0.005 to 0.008 second.

Oscillograph Diagrams—Fig. 77 is a so-called oscillograph diagram of the primary and secondary currents in a battery and coil, high tension ignition system, which was ob-

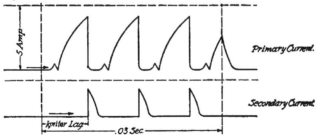

FIG. 77—OSCILLOGRAPH DIAGRAM OF PRIMARY AND SECONDARY CURRENTS.

tained by Prof. F. W. Springer. The dotted vertical line at the left represents the moment of closing the primary circuit. At this moment the current begins to flow in the primary circuit, increasing in value gradually. The resistance of this circuit is such that the current would attain a value of 5 amperes if the circuit remained closed long enough. However, when the current has attained a value of about 4 amperes the circuit is broken by the magnetic vibrator, and the current then falls off very rapidly.

It will be noticed that there is no current in the secondary circuit during the period the primary current increases, which is due to the fact that the e.m.f. induced in the secondary during this period is not sufficient to break down the resistance of the gap between the terminals of the spark plug. The spark occurs when the primary current decreases, and it is interesting to note that it attains its maximum value instantly, this having been confirmed by numerous oscillograph tests. The right-hand dotted vertical line represents the moment the primary circuit is broken by the timer.

With a magnetic vibrator system there is always a series of sparks in each cylinder when the engine runs at low speed, hence, if the first spark does not fire the charge, it may still be fired by one of the succeeding ones.

CHAPTER X

The Spark Plug

The spark plug, that member of the high tension system at which the ignition spark is produced, has to operate under very severe conditions, because in addition to being subjected to enormous electromotive forces it must also withstand very high temperatures. The plug consists essentially of three parts, viz., a steel shell which screws into the spark plug hole in the cylinder wall, and which forms one electrode of the plug; the central electrode, which consists of a steel rod, and an insulator between the central electrode and the shell. Among the insulators employed for spark plugs may be mentioned porcelain, steatite (soapstone), glass, lava and mica. The insulator must not only possess high insulating qualities, but it must also be capable of withstanding extremely high temperatures and considerable mechanical pressure. Porcelain is probably used most extensively, but special grades of porcelain of an exceedingly fine texture and not apt to crack under the effects of alternate heating and cooling, are required.

Insulators—Spark plugs may be divided into two classes, according to whether they are made in one piece or demountable. Where porcelain insulators are used they are generally demountable, the porcelain being provided with a circumferential flange at mid-length by which it is clamped between a shoulder on the inside of the shell and a screw bushing. (Fig. 78.) Copper-asbestos or asbestos cord gaskets are placed between the porcelain and metal parts to insure a gas-tight joint and to prevent injury to the insulator from undue mechanical pressure. Some other forms of insulators, like steatite, being less fragile, do not require these gaskets, and when these insulators are employed the plugs are generally made in one piece, the insulator being set in cement in the shell. Among the advantages of mica are that it is a first-class insulator and that it cannot be injured by any ordinary mechanical pressure, but it is subject to the objection that the insulator must be built up of discs and that

THE SPARK PLUG 97

even though it may be strongly compressed there still remain minute interstices into which conducting fluids may penetrate. To prevent a conducting path being formed through these interstices extending all the way from the outer shell to the central electrode, the latter is usually provided with a wrapping of sheet mica over which the mica discs are stripped. (Fig. 79.)

In porcelain plugs the central electrode is generally cemented into the porcelain, while formerly it was usually clamped into it, being provided with a head at the inner and a nut at the outer end. When cemented in place by means of a quick drying, heat resisting cement, the major portion

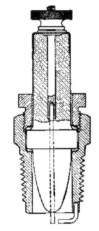
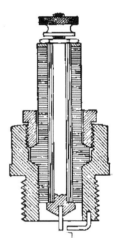

FIG. 78—ORDINARY PORCELAIN PLUG. FIG. 79—MICA INSULATED PLUG

of the length of the central electrode is made very thin, the reason being that the porcelain and the metal electrode do not expand and contract equally when subjected to changes in temperature, and when the electrode is very thin the porcelain is better able to withstand the strains due to this unequal expansion. In order to enable the binding screw of the central electrode to withstand any twisting strains that may be put upon it, it is brazed or welded to a cap which is anchored on the porcelain. Sometimes the central electrode is threaded over its whole length to insure a good grip.

Spark Points—In many plugs a wire of nickel, German silver or some other non-oxidizing metal is fastened into the

THE SPARK PLUG.

inner end of the shell and bent to come very close to the protruding portion of the central electrode. The spark gap is made about 1/32 inch for battery ignition and a little more than one-half that for magneto ignition, the obvious reason for the smaller length of gap for magneto ignition being that a lower e.m.f. is required to cause a spark to pass, hence the engine need not be cranked so fast to generate a spark for starting. Also, for magneto ignition, on account of the greater volume of the spark, the spark terminals are made heavier so they will not be burned away quickly by the heat of the spark.

Multiple Spark Points—There are many plugs on the market with multiple spark gaps. The spark, of course, always

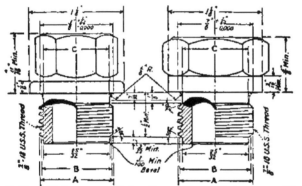

Small Hex., ⅞ in. across Flats. Large Hex., 1⅛ in. across Flats.
FIG. 80—S. A. E. STANDARD SPARK PLUG SHELLS.

	Nominal	Minimum	Maximum
A (outside diameter)	0.875	0.872	0.875
B (pitch diameter)	0.839	0.836	0.839
C (root diameter)	0.803	0.800	0.803

takes the easiest path, and only one of the gaps serves at any one time, but owing to the fact that the gap is likely to increase in length by reason of the burning away of the positive electrode by the spark, the different gaps will come into use in succession.

In one spark plug with three paths the three wires fastened into the shell have their ends opposite the central electrode flattened into a crescent, which is said to give a sheet of flame instead of a more compact spark, whereby the firing power is believed to be increased.

Spark Plug Threads—There are three standard forms of threads for spark plug shells. The S.A.E. standard is a straight thread ⅞ inch in diameter and having 18 threads to the inch. The metric standard has a straight thread 18 millimeters in diameter and a pitch of 1½ millimeters. Both of these straight thread plugs must be provided with a copper-asbestos gasket under their shoulder to prevent gas leakage. The third thread is a taper thread, the ½-inch standard pipe thread, which makes a gas-tight joint without a gasket. Plugs with taper threads must not be screwed too tight into the cylinder when the latter is hot, as it is then practically impossible to get them out, for later on the cylinder and plug will always be of substantially the same temperature and whether

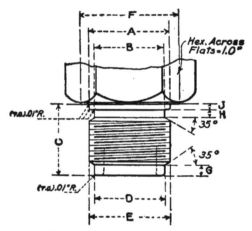

FIG. 81—S. A. E. STANDARD METRIC PLUG SHELL.
(18 Min. Thread Diameter, 1½ Min. Pitch.)
A, 0.706-0.710; B, 0.625-0.664; C, 0.625; D, 0.625 max.; E, 0.703-0.708; F, 15/16 min.; G, 3/32 min.; H, 5/64 max.; J, 3/64 max. Thread outside diameter, 0.703-0.708; pitch diameter, 0.664-0.669; root diameter, 0.620-0.625.

the plug is heated or the cylinder cooled, the result is the same—the plug becomes tighter in the cylinder. Fig. 80 shows the standard spark plug of the Society of Automotive Engineers in its two forms, only the parts outside the cylinder being different. The metric spark plug shell, as standardized by the S.A.E. for aircraft and motorcycle engines, is shown in Fig. 81. The standard thread for all spark plug terminals is No. 8-32 (0.164 in. diameter) A.S.M.E. standard.

Proper Length of Threaded Portion—It has always been a question as to how long the threaded part of the plug shell should be made, relative to the thickness of that part of the cylinder wall into which it is screwed. There are three possibilities which are illustrated in Fig. 82. At *A* the plug extends into the combustion chamber beyond the inner surface of the cylinder wall. This is objectionable because of the danger of the plug overheating. At *B* the plug does not extend entirely through the hole into which it screws, and a cavity is formed at its inner end. This cavity is apt to remain filled with dead gases when the engine is strongly throttled and thus cause misfiring. The proper length of the threaded portion of the plug is shown at *C*, where the end of the plug shell comes flush with the inside of the cylinder

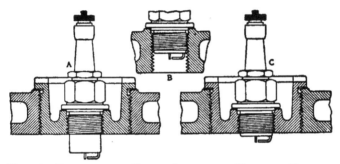

FIG. 82—IMPROPER AND PROPER LENGTH OF THREADED PORTION.

wall. As the wall thickness varies, spark plugs are made with several different lengths of thread, and the proper size should always be chosen.

Protected Porcelains—One of the most frequent troubles with plugs is that the exposed surface of the insulator inside the combustion chamber becomes coated with a layer of oil and carbon which forms a short circuit between the central electrode and the shell. One method of preventing this consists in making the shell of such form as to give protection to the insulator. That is, the shell is practically closed at its inner end, as shown in Fig. 83, so that projections of oil cannot reach it. There is also another advantage claimed for this form of construction, viz., that it accelerates flame propagation. At the bottom of the spark plug a small chamber is formed inside of which the spark occurs. The theory is that the charge inside this chamber, upon becoming ignited, greatly expands, and the burning mass is forced through the holes in the wall of the chamber at great speed, carrying combustion rapidly to the farthest part of the

THE SPARK PLUG.   101

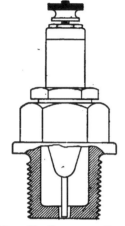

FIG. 83—PROTECTED PORCELAIN.

combustion chamber. The principle was first applied in conjunction with low tension ignition in the early Mercedes engines in which the make and break terminals were located in a separate chamber communicating with the combustion c h a m b e r through a restricted passage.

Non-Sooting R e c e s s—Another and more general method of preventing short circuits by sooting of the porcelain consists in forming the latter so there is a deep recess between it and the shell and sometimes also between it and the central electrode. This recessed construction is covered by a patent to Canfield under which many spark plug makers are licensed. The two constructions are illustrated in Fig. 84, A and B. Not only is the leakage path over the surface of the porcelain from the shell to the central electrode greatly lengthened, but it is also claimed that there is no deposition of carbon in the inner portions of the recess, because that space is permanently occupied by dead gas, consequently no combustion goes on there, and no carbon is formed. When a spark plug of this type is examined after having been in use for some time it will be found that while the surface of the porcelain at the mouth of the recess is strongly discolored by carbon accumulations, the surface near the bottom of the recess remains bright.

Location in Cylinder—In determining the best location of the spark plug in the cylinder wall several points must be taken into consideration. In the first place, the inflammation

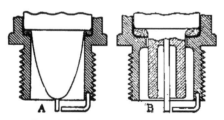

FIG. 84—RECESSED INNER ENDS.

of the gaseous charge, which starts from the spark gap of the plug and is thence propagated throughout the mass of the charge, should be completed in the shortest possible time, and this consideration would lead us to place the spark plug so that the spark gap came substantially in the center of the combustion chamber. This would be impractical, however, for another reason, because it would be impossible to effectively cool the spark points and porcelain. So far as cooling is concerned, a good place for the spark plug is in the filler plug over the inlet valve or in the cylinder wall over the passage from the inlet valve chamber to the cylinder, where the inner end of the plug is cooled by the incoming charge. This location is also advantageous for another reason, namely, that when the engine is strongly throttled and the amount of fresh charge admitted is small in comparison with the dead gases remaining in the cylinder from the previous explosion, the purest charge is most likely to be found around the inlet valve, and hence, if the plug is located there, ignition is most reliable.

When two sets of spark plugs are employed, one set is sometimes screwed into the plugs over the exhaust valves, but this is not a good location, as in addition to poor cooling and diluteness of the mixture in the exhaust valve chamber at low throttle, the exhaust gases passing by the plug carry oil and carbon and the latter is apt to become dirtied and short circuited in a short time. In a two-stroke motor where the charge transferred from the precompression chamber to the combustion chamber carries the oil for cylinder lubrication, the spark plug should not be located in the path of the incoming charge.

Spark Plug Faults—Probably the most common fault of spark plugs is "sooting"; that is, the surface of the insulator which is exposed to the action of the flame becomes covered with carbon which forms a short circuiting path aross the spark gap and thus prevents the production of a spark. This can be remedied by cleaning the surfaces affected by means of a small, stiff brush dipped in gasoline, taking the plug apart for the purpose. The porcelain insulators of spark plugs will sometimes crack, owing to sudden local changes in temperature which may be caused by water being splashed on them while hot. Their insulation is then insufficient to hold the spark, and either a new porcelain or a new plug is needed. Care must be taken not to expose the spark plugs to excessive temperature, as by greatly overheating the engine, and when decarbonizing an engine cylinder by the oxygen method, no good spark plug should be left in the cylinder. The decarbonizing tool is usually introduced through one spark plug hole, and if the cylinder ordinarily carries two plugs the other one should be replaced by a dummy while the process is being applied.

It is important that the distance between the spark points be maintained, and if the gap has become too great owing to burning away of the spark terminals or accidental bending, the terminals should be dressed up by means of a fine file and bent so as to be the proper distance apart. This is particularly important where magneto ignition alone is used and the engine has to be cranked by hand. Some manufacturers of plugs furnish a gauge consisting of a small piece of sheet metal of a thickness equal to the proper width of the gap. When the central electrode consists of a thin wire which is joined to the binding post, it will sometimes come loose from the binding screw and form an extra spark gap, which may result in missing at low engine speeds, if current is supplied by a magneto. The fault can be detected by feeling of the inner end of the central electrode, and if it exists a new plug should be substituted.

Occasionally spark plugs will become short circuited by particles of carbon or other conducting material lodging between the spark points. Such a fault, of course, is readily corrected.

Auxiliary Spark Gap—When a spark plug has become short circuited from sooting and in consequence fails to fire the charge, it may be caused to do so again by providing a second gap in the high tension circuit somewhere outside the cylinder, about equal in length to the gap between the spark plug terminals. Of course, under ordinary conditions it is much preferable to clean the plug or replace it by a new one, but fifteen years ago when the oiling systems of engines were rather imperfect and "sooted plugs" were a chronic ailment, the "auxiliary spark gap" did good service. It may not be readily apparent why the introduction of an additional spark gap in series with the regular one should cause a spark to pass, but the explanation is as follows:

The secondary e.m.f. does not attain its maximum value instantly but builds up gradually though exceedingly rapidly. Ordinarily it will continue to grow until it is strong enough to break down the resistance of the spark gap. However, if the spark gap is short circuited by a layer of carbon on the porcelain, current will begin to flow in the secondary circuit as soon as the e.m.f. begins to build up. When there is no current flowing the whole of the e.m.f. induced in the secondary winding of the coil is impressed upon the spark plug, but when a current flows, a great deal of this e.m.f. is used up in overcoming the resistance of the secondary winding and what remains is not sufficient to break down the resistance of the spark plug air gap. When a second air gap is introduced into the circuit, as shown at A, Fig. 85, no current can flow though the plug may be heavily sooted, and the e.m.f. then builds up in the same way as ordinarily, until it is sufficient to break down the combined resistance of the two air gaps, which, since the second gap is under atmospheric pressure, is only about 20 per cent greater than that of the spark plug gap alone. It might be supposed that the current, after passing the auxiliary gap, would follow the leakage path

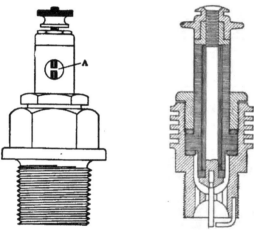

FIG. 85—AUXILIARY SPARK GAP PLUG.   FIG. 86—PITTSFIELD AIRCRAFT PLUG.

around the plug gap, but the sudden impulse of current involved in a spark will always follow the straightest possible path even though it offers more resistance, hence it will jump between the spark terminals at the gap between them, rather than follow the leakage path. If we make use of an hydraulic analogy we may say that the auxiliary gap acts like a dam which holds the current back until the water level (e.m.f.) rises sufficiently high to break through the dam (jump the gap) and the current thereby attains such momentum that rather than flow through a small by-pass (leakage path) it will continue straight on and break down the second dam (spark plug gap).

Originally the auxiliary gap was furnished by a separate device connected in the high tension circuit in series with the spark plug; later it became customary to incorporate it in the spark plug itself as shown in Fig. 85. Such a plug also gives a visible indication of the spark. These auxiliary spark gap plugs, however, never achieved any great popularity, and the same effect is now obtained by contact-less distributers, as will be explained later on.

Aircraft Plugs—Reference was made above to the severe conditions under which the spark plugs of automobile engines are compelled to work, but the conditions are far worse on aircraft engines, because the compression ratio is higher in these engines and because they work under full open throttle all the time. Where the service is so severe the plug cannot be screwed into the valve plug but must be screwed directly into the cylinder wall, where the cooling water can be brought close to it. Also, it has been found necessary to design special plugs with extra cooling means to keep the temperature of the plug down and prevent pre-ignition of the charge. Similar conditions obtain in kerosene burning tractor engines because of the high load factor in farm tractors and the extra heating of the engine with kerosene fuel, at least with most present kerosene vaporizing systems. Air-cooled motor cycle engines also are rather hard on spark plugs.

In Fig. 86 is shown the Pittsfield aircraft plug, which has mica insulation. It differs slightly from the ordinary mica plug in that the shoulders of the mica core are square. The lower end of the shell is provided with a perforated baffle plate which tends to keep oil away from the mica. The central electrode stem is made of brass or copper, of large diameter, to afford plenty heat transmission capacity, and the electrode proper is swedged into the stem. The shell is finned to provide greater heat radiating capacity, and there is also a fin at the top of the stem to increase the radiation of heat from the stem and electrode. The top of this finned portion is slightly countersunk and the stem is riveted into same, thereby reducing the possibility of leakage past the threads on the stem. This finned portion is necked to take a slip terminal. The fins on the shell are made hexagonal to take a wrench.

Porcelain insulated plugs are also being successfully used on aircraft engines, but it was found necessary to greatly improve the grade of porcelain over that commonly used for spark plugs, especially with a view to increasing the strength at high temperatures. The glaze also had to be improved to bring its coefficient of heat expansion close to that of the porcelain itself, as otherwise it will flake off as soon as the plug is heated up. A writer in *Automotive Industries* (Jan. 3, 1918), who investigated the aircraft spark plug situation for a foreign Government, sums up as follows the requirements that must be met in order to make a spark plug satisfactory for this class of work:

"All materials used for the main parts of spark plugs must have a low coefficient of heat expansion. This coefficient must not be the same for all parts but inversely proportional to their heat conducting properties. For instance, the heat-expansion coefficient of the central electrode should be less than that of the porcelain, because the central electrode reaches a higher temperature. Where it is impossible to fulfill this requirement it is necessary to make such provisions as will allow the expansion of a certain part to take place without injuring the other parts of the plug.

"The central electrode must be well cooled, preferably by means of fins at the upper end, and only a small part of the electrode must be exposed to the heat of the combustion chamber. That is, the extension of the electrode beyond the lower end of the insulator must be short and of small diameter— of sufficient diameter only to prevent warpage under heat. All joints in the central electrode should be of welded or similar type, so as to ensure proper heat conductivity. The ground electrode must be short and rigidly held, and should preferably be of the single point type. The lower end of the shell should be given such a shape as to protect the insulator from oil and carbon, unless the material used is of sufficient heat-resisting quality to withstand the heat of the explosion and permit the carbon to be burned off. If cement is used, this must not soften or deteriorate at temperatures somewhat above service temperatures."

CHAPTER XI

Vibrator Coil Ignition for Multi-cylinder Motors

What has been said so far regarding battery and coil high tension ignition has application chiefly to ignition of single cylinder engines. When the ignition of a multi-cylinder engine is considered, we may use a separate vibrator coil for each cylinder and a timer with as many contact segments spaced around its circumference as there are cylinders to be ignited. This was the first solution of the multi-cylinder high tension ignition problem and has been most extensively employed. The individual coils in their wooden boxes are generally arranged in a hardwood box mounted on the dash, in such a manner that the terminals on the outside of the individual coil automatically make contact with corresponding metal parts in the bottom or on the sides of the coil box. In the case of the two terminals carrying primary current it is essential that good metallic contact be made at all times, and these are preferably made of such a form that as the

FIG. 87—QUADRUPLE VIBRATOR COIL IN BOX WITH COVER REMOVED.

VIBRATOR COIL IGNITION

coil is forced down into the box its contacts wedge between or against those in the box. In the case of the terminal which carries high tension current only, a good connection is not essential, as the high e.m.f. will easily break down whatever resistance may be offered by the joint. The ignition switch is usually mounted on the front of the coil.

Multiple Vibrator Coil Systems—Fig. 88 shows a wiring diagram for a four-cylinder, battery and coil ignition system employing multiple vibrator coils. In this connection it must be remembered that in a four-cylinder engine the cylinders cannot be fired in regular order, starting from one end, but

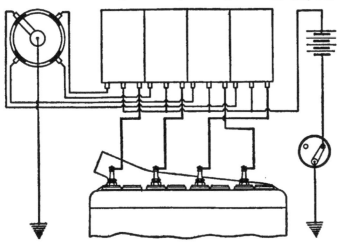

FIG. 88—WIRING DIAGRAM OF MULTIPLE VIBRATOR COIL SYSTEM.

must be fired in one or the other of the following orders (starting from the front):

1—2—4—3.   1—3—4—2.

It should be remembered that each coil has three terminals, and in Fig. 88 the terminal on the left is the low tension terminal which connects to one of the terminals of the timer; the centre terminal is the ground terminal which is grounded through the battery and switch, and the right hand terminal is the high tension terminal, which is connected to one of the spark plugs. Of course, the grounding terminals of all four coils are connected together. The circuit of the individual coil is the same as already described in connection with Fig. 75.

Buckproof Coils—When a number of coils are placed in close proximity of each other, each coil has a rather strong inductive effect on the adjacent coils. In Fig. 89 are shown the cores of two adjacent coils in their true proportions. It will readily be seen that when one of the coils is energized a large proportion of the magnetic lines of force through its core will return through the core of the adjacent coil, because that, together with the intervening air space constitutes a path of less magnetic resistance than the return path all the way through air. But the establishment of a magnetic field of force through this adjacent coil may cause an e.m.f. to be induced in the secondary of that coil strong enough to produce a spark at the plug connected to that coil, and we then are likely to have a case of back firing. When one cylinder is at

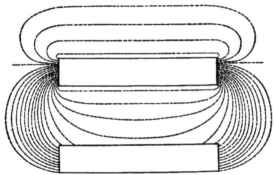

FIG. 89—ILLUSTRATING MUTUAL INDUCTIVE EFFECT BETWEEN ADJACENT COILS.

the beginning of the power stroke, where ignition in it properly occurs, the cylinder whose coil is located adjacent to that of the former, may be at the beginning of the compression stroke, when it is full of combustible charge and when, on account of the absence of compression, a comparatively small e.m.f. is sufficient to produce a firing spark.

This may be remedied by placing a shield of some highly (electrically) conductive material between adjacent coils. When magnetic impulses pass through these shields, local currents are set up in the shield which oppose the variation of the magnetic flux through them and therefore hold the flux down to a very low value. These currents are known as eddy currents.

Single Vibrator Coil and Distributer System—A spark coil is a somewhat expensive piece of apparatus, and the thought,

VIBRATOR COIL IGNITION

therefore, occurred that a single coil might be used for sparking any number of cylinders. In that case, instead of closing the primary circuit once during every two revolutions of the engine crankshaft, it must be closed during this period a number of times equal to the number of cylinders to be sparked. This will produce in the secondary winding the necessary number of impulses to spark all of the cylinders. It is, of course, evident, that in this case the secondary winding of the coil cannot be permanently connected with the spark plugs, but must be connected to the different spark plugs in rotation, which is accomplished by means of a device known as a high tension distributer. In principle the high tension distributer

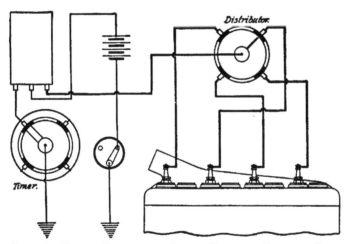

Fig. 90—Wiring Diagram of Single Vibrator Coil and Distributer System.

is similar to a multi-cylinder timer, but in details of construction it is different, as it must be insulated to withstand very high e.m.f.s., while, on the other hand, no very good electrical contact is necessary and the rotating arm and segments need not even come into actual contact.

Fig. 90 is a diagram of connections of a single coil, high tension distributer ignition system for a four-cylinder motor. It will be noted that the connections of the primary circuit are similar to those of a four-coil system. All four contact segments of the timer are metallically joined together and from them a wire is run to the low tension terminal of the coil. The ground terminal of the coil is connected to ground

VIBRATOR COIL IGNITION

through the battery and switch. From the high tension terminal of the coil a wire leads to the central rotating terminal of the distributer and from the four distributer segments cables are run to the four spark plugs respectively.

In this system the timer and distributer must both be driven so as to make one revolution for every cycle of the engine and consequently the two devices are usually, made in a single unit. Fig. 92 shows a typical construction.

Combined Timer and Distributer—Fig. 91 shows a typical design of a combined timer and high tension distributer, due to the Dayton Electrical Laboratories Company. The timer forms the lower part and the distributer the upper of the in-

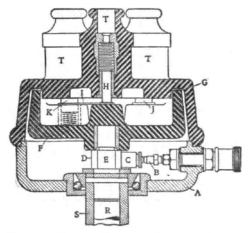

FIG. 91—VERTICAL SECTION THROUGH COMBINED TIMER AND DISTRIBUTER (DELCO).

strument. The timer comprises a cam D on a short shaft E which connects through the sleeve S with the vertical shaft R driven from the engine; also a vibrator spring C actuated by cam D, and a stationary contact B carried in the metal housing but insulated from it. Another view of this timer construction which brings out its details more clearly may be seen in Fig. 98.

The high tension distributer consists of two moldings of insulating material, viz., the stationary part G, which forms the cap of the housing and carries the high tension terminals TTT, and the rotating part F, also known as the rotor, which is secured to the top end of the timer shaft. High tension current from the secondary winding of the coil enters the high tension distributer through the central terminal *T* and flows through pin *H* to the metal strip *I* secured to the rotor. In the stationary part *G* are carried four or six contact pins, according to whether the instrument is to be used on a four or a six-cylinder motor, each of which connects to the spark plug of one cylinder. The rotor carries two contact brushes, of which brush *K* serves to connect the central pin *H* successively with each of the four or six terminals connecting to the spark plugs. Brush *J* serves to connect the spark plug cable of the cylinder next in order to fire to ground, so that any static charge which may be induced in the high tension cable leading to the plug will be conducted to ground and not cause a discharge at the spark plug of that cylinder in

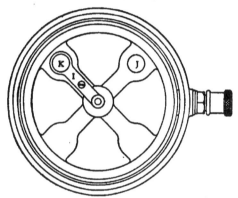

FIG. 92—PLAN VIEW OF DELCO DISTRIBUTER WITH COVER REMOVED.

which, if the spark lever is set for early ignition, the suction stroke is still uncompleted and the entering charge might be ignited which would cause a back fire through the carburetor. Prevention of improperly timed explosions due to static induction between adjacent cables will be further dealt with under the heading of wiring. The feature of grounding the cable leading to the cylinder next in order to fire through the distributer is special to the design of timer-distributer shown and is not found in all such instruments.

Master Vibrator—To the multiple vibrator coil system the objection is often made that, owing to differences in the actions of the vibrators, which may be due either to differences in construction or to differences in adjustment, the sparks in the different cylinders will not be timed alike

VIBRATOR COIL IGNITION

or will not be of equal power, which tends to unbalance the engine. It is, of course, almost impossible to adjust four or six vibrators exactly alike, and this has led to the use of a single vibrator, called a master vibrator, with a number of plain coils. The master vibrator comprises a small coil included in the primary circuit, whose only object is to energize the magnetic core for operating the vibrator. A timer connects the primaries of the different plain or non-vibrator coils in circuit in rotation and the secondary of each coil is permanently connected to its spark plug. Fig. 93 is a wiring diagram of a master vibrator system for a four-cylinder engine. This system is used mainly as a substitute on cars

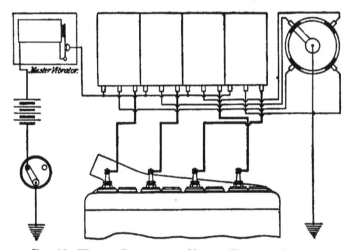

FIG. 93—WIRING DIAGRAM OF MASTER VIBRATOR SYSTEM.

originally fitted with multiple vibrator coils. The magnetic vibrators of the original coils are then short circuited by means of a wire, and the coils are used as plain coils in the master vibrator system.

In the foregoing the four-cylinder engine has been used as a representative of the multi-cylinder type. It will be understood that in the case of a six-cylinder engine the timer has six equally spaced contact segments and in the case of an eight-cylinder engine, eight.

Ignition Timing—The contact segments of a roller contact timer usually extend over an arc of about 40 degrees, and as the roller makes and breaks contact when its axis is in a plane with the timer axis and the edge of the segment, contact is maintained for 40 degrees of camshaft motion and (in a four-stroke engine) 80 degrees of crankshaft motion. The timer must be so set on its shaft that when in the maximum advance position the first spark in the cylinder at very high speed will occur not later than on dead centre. Let us assume that the vibrator has a periodicity of 250 per second, then the igniter lag is 0.004 second. Suppose the engine is to run at a maximum speed of 1500 r.p.m. or 25 r.p.s. Then the time of one revolution (360 degrees) of the crankshaft is 0.040 second, and if the spark at that speed is to occur on dead centre when the timer must make contact $\dfrac{0.040}{0.004} = 1/10$ revolution of the crankshaft ahead of the dead centre position. One-tenth of a revolution or 36 degrees corresponds to 19/32-inch piston travel to dead centre with a 5-inch stroke and a connecting rod length equal to twice the stroke. It is advisable to make the maximum lead of the timer slightly greater than this, say ⅝-¾ inch of piston travel for a stroke of 5 inches, and in proportion for other strokes.

If, then, it be required to set the timer on its shaft so that it will be timed properly, first slip it over the shaft and connect it up to the spark lever on the steering wheel. Then set the spark lever in the "maximum advance" position and connect the terminals of the timer to the spark coils. The rotating member of the timer is still loose on its shaft. Now turn the engine until the piston in cylinder No. 1 is say ¾ inch from the top end of the compression stroke (the upstroke following the closing of the inlet valve). While in this position turn the movable part of the timer on its shaft in its normal direction of rotation until the roller or equivalent contact member begins to make contact with the segment joined up to the coil whose secondary connects with the spark plug in cylinder No. 1. While in this position secure the movable part of the timer to its shaft. If all the electrical connections have been properly remade the ignition is then properly timed for all of the cylinders. When taking off the ignition wires and cables it is well always to attach small pieces of paper with like marks on them to the end of the cable and the terminal or binding post to which it connects, respectively, so no difficulty will be encountered in remaking the connections.

Location of Troubles—If one cylinder of a multi-cylinder engine misses and it is desired to determine which one it is, this can readily be done by making sure that no spark can be produced in that cylinder and noting the result. To this end it is only necessary to short circuit the spark plug by placing a screw driver or similar tool so it connects the spark plug cap or binding nut with the outer shell of the plug or any part of the engine. If there is no change in the operation of the engine then it is this cylinder which has been giving the trouble. On the other hand if the spark plug short-circuited is not in the missing cylinder, then two cylinders will miss and the result will immediately be noticeable in the speed of the engine and the greater number of missing exhaust noises.

One advantage of a magnetic vibrator system is that the "buzz" of the vibrator when the circuit is closed through the timer gives an indication of the fact that the primary circuit is all right—that the batteries are giving current and that there is no break anywhere in the circuit. Of course, sometimes when the batteries are nearly run down they may give still enough current to operate the vibrator but not enough to produce a spark of sufficient energy to fire the charge. But the buzz of the vibrator will then be comparatively faint, which will be at once detected by the trained ear.

Adjustment of Vibrator—The contact points of the vibrator wear away in use, and this necessitates occasional readjustment. The manner of adjustment varies somewhat with the construction of the vibrator. Many vibrators have two adjusting screws or nuts (Fig. 94), one being a screw with a platinum tip which forms one of the contact points. The adjustment of the latter screw determines the distance of the armature from the end of the magnetic core and the adjustment of the other screw or nut determines the pressure between contact points. It is obvious that the greater the pressure between contact points the more magnetic pull is required to draw them apart and, consequently, the greater the value to which the primary current will rise before the circuit is broken. On the other hand, if the contact screw is adjusted to bring the armature nearer to the core, then the magnetic effect of the core on the armature for a given primary current will be increased.

Most manufacturers of coils state the current which their coils should consume to operate properly on engines with different compressions, and if a small ammeter is available the best plan is to connect it in circuit and change the adjustment until the desired current consumption is obtained. In this connection it is well to remember that if the coil is so adjusted that the contacts are drawn apart too soon, no effective spark will be produced, while if they are so adjusted

116   VIBRATOR COIL IGNITION

that the points press together very hard, then too much current is consumed. Therefore, without an ammeter, the best method of adjustment probably is to "loosen" the vibrator while the engine is running until the particular cylinder begins to miss, and then slightly turn the adjusting nut the other way. It may be stated that the current consumption of vibrator coils varies from $\frac{1}{2}$ to 2 amperes.

Dressing Up Contact Points—Owing to the wear on them the contact points in the course of time become rough and then offer considerable resistance to the passage of the current. To remedy this they must be dressed up with a fine

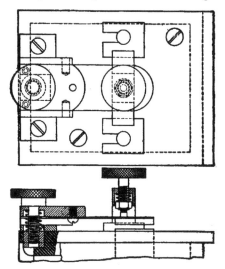

FIG. 94—MAGNETIC VIBRATOR WITH TWO ADJUSTING SCREWS.

file. It is the positive contact, the one from which the current passes over to the other, that is affected most, and the contact point which can most readily be removed from the coil (the contact screw if one is used) should always be made the positive one. In dressing up the point of a contact screw care must be taken to get the surface flat and perpendicular to the axis, so there may be contact over the whole surface, and this is best accomplished by making a jig or holder consisting of a flat piece of metal about $\frac{3}{8}$-inch thick, which is drilled and tapped with a thread corresponding to that of the contact screw so that the latter fits into it tightly. It is then screwed into the jig so its point barely projects, and is filed off flush with the surface of the jig. Of course, the hole through the jig must be drilled and tapped square with the surface of the same.

Care of Coils—Coils must not be operated on a higher voltage than they are intended for, and this becomes especially dangerous when the secondary cable should accidentally be disconnected from the plúg, or the secondary circuit be interrupted in some other way. Ordinarily the e.m.f. in the secondary circuit is limited by the distance between spark points and by the compression, but if the secondary cable should be disconnected, the secondary e.m.f. would increase with the primary e.m.f. and there would be danger of breaking down the insulation of the secondary. To obviate the production of excessive e.m.f.'s some coil makers bring wires connected to opposite ends of the second winding to within a distance of ½ inch or so of each other. Then, when the secondary voltage attains that necessary to jump a gap of ½ inch in the atmosphere a spark will pass at this "safety spark gap" and there will be no further increase in voltage. A failure of the insulation of the secondary such as might be caused by excess secondary voltage is fatal. That is, it is better to get a new coil than to try making a repair. Fortunately, the way coils are now made, if they are not grossly abused, such failures are exceedingly rare.

Owing to the fact that the insulating wax used in coils melts at a moderately high temperature, coils must never be placed where they are close to parts radiating a great deal of heat, like the exhaust pipe. That would be likely to cause the insulating wax to run and thus impair the insulation.

Locating Faults—A common method of determining whether all of the plugs spark is to unscrew them from the cylinders, lay them on top of same with the high tension cables connected and turn the engine over by hand. While this method may show the entire absence of a spark in one or more cylinders, it is by no means a dependable test, for the reason that to produce a spark under compression in the engine cylinder requires several times the e.m.f. required to produce a spark of equal length in the atmosphere. Spark plug testing devices have been made which consisted of a little chamber with a window into the wall of which the spark plug could be screwed and in which an air pressure could be pumped up by means of a tire pump. However, these never came into practical use, as with multi-cylinder engines the best plan is to test the plug while in place on the cylinder.

Where a high tension distributer is used it must be located where it is not likely to become wet and dirty, as the least impairment of its insulation interferes with its proper operation. The timer needs a little oil occasionally and must also be kept clean so that good electric contact may be assured and there may be no leakage of current to a contact segment before the roller or other movable contact member reaches it.

CHAPTER XII

Modern Battery Systems

The system of ignition described in the early part of Chapter IX is the De Dion, the first high tension system to come into extensive use. It was abandoned in favor of magnetic vibrator systems because its mechanical vibrator was too delicate and unreliable. At later periods in the development of ignition equipment various improved types of mechanical vibrators were evolved, all of which can be divided into two classes, the open circuit and the closed circuit. The open circuit vibrator owes its development to the demand for extraordinary current economy when engine cylinders began to multiply and dry cells still served as sources of ignition current, and the closed circuit type made its appearance subsequent to the adoption of electric starting and lighting systems, when the presence of a dynamo generator on the car seemed to make a magneto generator for ignition superfluous, provided the spark generating and distributing devices could be made equal in reliability to the corresponding parts of the high tension magneto. High current economy was not essential but the interrupter must operate reliably at high engine speeds, as the period referred to marked the introduction of the high speed gasoline engine for automobile service. Generally speaking, in an open circuit system the time the interrupter contacts remain together is independent of the engine speed, whereas in a closed circuit system this time varies inversely as the engine speed.

Single Spark Vs. a Shower of Sparks—It was at one time believed that there was a substantial advantage in having a series of sparks occur in the cylinder at the beginning of each power stroke, so that in case the charge were not fired by the first spark there would still be a chance that it would be fired by one of the following ones. That this will occur is undeniable, but the benefit to be derived therefrom, except in cranking the motor, is very questionable. At high engine speeds ignition by the second spark would come so late in the stroke that the explosion would add little to the power developed, but, on the other hand, it would tend to increase heating of the cylinders. Moreover, if the first spark

FIG. 95—ATWATER KENT INTERRUPTER SHOWN IN DIFFERENT POSITIONS.

missed frequently, the engine would run very irregularly, and, in such a case, the best plan certainly would be to investigate the cause of this missing and remedy it immediately. Another thing to be taken into consideration is that the series of sparks is not a feature directly aimed at in the design, but is rather incidental, for in order to insure the production of a good-sized spark at high engine speed the timer contact segments must be made to cover a certain minimum arc of the circumference, and with such an arc of contact a series of sparks will be produced at low speeds. There is no more need for a series of sparks at low speed than at high speed, and the fact that it is obtained only at low engine speeds seems to prove that it is incidental and not specially aimed at by the designer.

About the time when four cylinder engines first came into use the development of the so-called single spark type of interrupter was begun. In some of these interrupters the duration of contact is not affected by the speed of the engine, and it is obvious that these permit of the highest current economy. The writer believes that A. Atwater Kent was the pioneer in this field of development, and the Atwater Kent device will here be described as typical of this class of interrupters.

Atwater Kent Open Circuit Interrupter—Referring to Fig. 95, the Atwater Kent interrupter (or contact maker, as it is called by its manufacturer) comprises a notched shaft A, which is driven from the engine crankshaft at one-half crankshaft speed in four stroke, and at crankshaft speed in two stroke engines. This shaft has as many notches on its circumference as there are cylinders to be fired (in this case six). With the notches on shaft A engages a hooked sliding part B, called the lifter. As shaft A rotates the saw teeth on it engage the hook on lifter B, and draw this part in the direction of its length, against the tension of coiled spring C When a certain position is reached, owing to the rotary motion of shaft A, lifter B is disengaged from the notch in the shaft, and it flies back under the tension of spring C. In doing so its head strikes a blow to "hammer" D—a swinging lever, whose motion in both directions is limited—and the hammer communicates the blow to the contact spring E. The latter, it will be noted, is of peculiar construction, comprising one substantially straight spring blade and a curved blade riveted to the former and having its outer free end bent into the form of a C curve, which extends behind the end of the straight blade. The latter carries the movable contact point, which is adapted to contact with a stationary contact point carried on the contact screw F, opposite it. Ordinarily the straight spring blade is held under tension by the hook of the curved spring blade, and the two contact points are separated. When the curved blade receives a blow from hammer

122   MODERN BATTERY SYSTEMS

D, the points are brought into contact. This limits the motion of the straight blade. The curved blade, however, is thrown over farther by the impact, and its hook leaves the end of the straight blade. After having reached the limit of its motion it flies back and strikes the end of the straight blade a blow which insures a very sudden break of the circuit.

Fig. 95 shows successive positions of the parts. In sub-figure A a notch in shaft A has engaged B and is drawing it against the tension of spring C. In sub-figure B the lifter is released. In sub-figure C the lifter is riding back over the rounded portion of the shaft and striking hammer D, which in turn pushes E for a brief instant against F. The return of B to its final position, as shown in sub-figure D, is so quick that the eye cannot follow the movement of D and E, which seem to remain stationary.

Important features of this device are the brevity of contact, its constant duration regardless of engine speed, and the quickness of the break. These tend to reduce heating and arcing at the contacts to the minimum, while still giving a strong current impulse in the coil. To further increase the life of the contacts a special type of switch is used with this system, which reverses the flow of current every time it is turned. The makers state that the contacts need adjustment only about once in 10,000 miles of average use, and replacement only once in 50,000 miles. Adjustment is made by removing one or more thin washers under the head of screw F.

All the moving parts, A, B, and D, are of steel, accurately made and glass-hard, and B and D are extremely light. The forces involved are therefore minute and the wear is stated to be negligible.

The contact points of this interrupter are made of tungsten, which is claimed to be more durable than either platinum or iridium.

It is impossible with this system to run down a battery by leaving the switch closed, as the lifter B only produces contact while it is returning, as in sub-figure C, and when the parts are at rest in any position of the shaft contact is definitely broken, regardless of the switch.

Closed Circuit Type Interrupter—In the closed circuit type interrupter the contacts are normally together, and are separated positively when the spark is to be produced. Fig. 96 shows such a device as manufactured by the Connecticut Telephone & Electric Company. A single armed lever A, made of sheet metal, so as to have a comparatively small moment of inertia, and of substantially semi-circular form, carries the movable contact point B. The latter is normally pressed against the stationary contact point C by the spring D. At the middle of its length lever A carries a fibre roller E, which is acted upon by a four-lobed cam G, carried on the shaft of the interrupter. There are two binding posts on the outside of the instrument, one of which connects with stationary contact point C and the other through a metal strip F with spring D and hence with movable contact point B.

One advantage of such an interrupter is that it has no lag. The primary circuit is broken and the spark in the cylinder occurs always exactly at the same point in the cycle of the engine. It, therefore, would be possible to do away with the ignition control mechanism where this type of interrupter is used, nevertheless, it is common to use hand-timing mechanism also with interrupters of this class, for the reason that owing

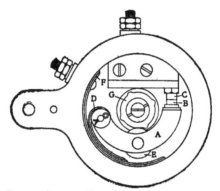

FIG. 96—CLOSED CIRCUIT TYPE INTERRUPTER (CONNECTICUT).

to the time consumed in propagating the flame through the whole mass of the charge the spark must occur earlier at high engine speeds than at low. A system employing a vibrator of this type is not very economical of current, as the primary circuit remains closed except for a short period when the contact points of the interrupter are separated, but at present this is of little consequence, as nearly all cars carry electric generating equipment of adequate capacity. The interrupter lever is insulated from the base of the interrupter, and, as the coil furnished with this interrupter has its primary winding not connected to the secondary, the primary circuit may have an insulated return.

Automatic Switch and Battery Savers—With a magnetic vibrator ignition system, if the driver should stop the engine by closing the throttle and then forget to open the switch, if the circuit happened to be closed in the timer, he would be warned that current was being wasted by the buzz of one of the vibrators. No such warning is given in the case of an ignition system with a magneto type interrupter, and the battery is likely to be drained completely or the coil injured by overheating unless provisions are made in the system to prevent this. In the Connecticut battery ignition system a thermostatic switch is used which opens the circuit in case the operator forgets it or when, on account of a derangement of the lighting and starting circuits, an excessive current flows through the ignition circuit.

Referring to Fig. 97, the current from the battery enters at A or B, and when the switch is closed at either C or D,

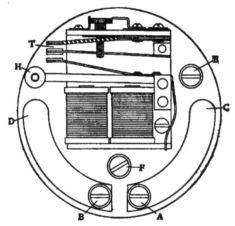

FIG. 97—CONNECTICUT AUTOMATIC SWITCH.

it flows to the thermostat, through the heater tape which is coiled around the thermostat, and then to post E. When the circuit is not being interrupted, that is, when the motor is not running, a much stronger current flows than in the regular operation of the system. This excessive current causes the thermostatic arm to bend and make contact at point T, thereby allowing the current to flow through the magnets and return to post F. This sets up a vibrator or hammer action in hammer H which drives away the retaining mechanism and unlocks the switch.

Other manufacturers connect a coil of iron wire in circuit with the primary winding of the induction coil. When the current flows continuously this iron wire heats up much more than in ordinary operation, and iron has the property that when raised to a certain temperature its electrical resistance suddenly increases greatly. Thus after this temperature has been attained the current which can flow through the circuit is only small and the battery charge is conserved.

Delco Relay—A device by means of which either a series of sparks or a single spark can be produced was manufactured until recently by the Dayton Electrical Laboratories Company, makers of the Delco electric systems. Since this device is fitted to thousands of cars now running, a description of same may be of practical interest, even though the device is no

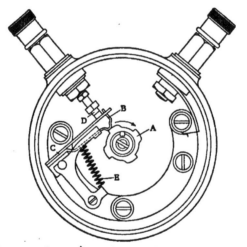

FIG. 98—DELCO OPEN CIRCUIT TYPE INTERRUPTER.

longer being manufactured. It was claimed by the manufacturers that a series of sparks was beneficial for starting, as the gasoline is likely to condense on the inner parts of the plug, and thus make it difficult to explode the charge, but that the heat of a series of sparks would vaporize this condensed fuel and thus facilitate starting. The advantages of a single spark in regular operation are greater current economy and reduced wear of the vibrator points.

The Delco ignition system comprises a form of vibrator known as a relay, which is connected in series with a mechanical interrupter or timer. This timer, Fig. 98, is of somewhat unusual form, and may first be briefly described. On the timer shaft there is a cam, A, with as many lobes as there are cylinders to be fired. This cam works on a spring B, bent as shown. A straight spring blade C carries the movable contact point, and its end extends through a slot in the end of the bent spring B. The movable contact point is normally held away from the stationary contact point D by the tension of coiled spring E, which places spring C under a slight tension. When the cam lobe comes opposite the bump in spring B, it deflects the latter and allows the contact points to come in contact. As it passes the bump, spring B is drawn back by coiled spring E, its end hits spring C, and the contact is sharply broken. It is obvious that with this device the duration of contact varies with the speed of the engine.

The Delco relay is shown in Fig. 99. It consists of an electro-magnet with two windings, one of comparatively coarse, and the other of very fine wire. Both windings are designed to be connected to the same source of current. The one made up of coarse wire produces the greatest magnetic effect on the armature, exerting a sufficiently strong pull on it when in its position of rest to draw it toward the end of the magnet core. The fine wire winding is designed to produce a smaller magnetic effect, not sufficient to draw the armature from its position of rest toward the core, but sufficient to hold it to the core once it is there. The condenser R is made up in cylindrical form and located adjacent to the electro-magnet. The magnet and its armature are of unusual construction. The central magnetic core extends some distance below the coils, and an L shaped piece N of soft iron extends up the side of the coil, to form, together with the magnet core and the bell crank form of armature A, a substantially closed magnetic circuit. The armature is pivoted to the top of the piece N, and its downwardly extending arm carries a heel, B, which strikes the spring blade C, carrying one of the contact points, the other being carried by another blade, D. An additional spring blade E, anchored together with the other two, insures that the two contact points are firmly pressed together when the armature A is not attracted by the magnet core.

In connection with this system a three point switch is used, having an "off" position, a "starting" position and a "running" position. In starting the connections are as shown in Fig. 100 in full lines, the fine wire winding of the relay remaining unenergized. When the circuit is closed by the timer, current from the battery will flow through the primary winding of the so-called plain induction coil and through the coarse winding of the relay. As soon as the relay becomes energized its armature begins to vibrate and the contacts which it operates make and break the circuit in rapid succession. The result of the successive interruptions of the primary current is a series of high tension impulses in the secondary winding of the induction coil, which produce a series of sparks at the spark plug to which the secondary happens to be connected at the moment by the high tension distributer. This continues until the primary circuit is broken in the timer.

At high engine speeds it is preferable to produce only a single spark per explosion, for reasons already stated. The switch handle is then moved to the "running" position, and the circuits are completed, as shown in Fig. 100 by dotted lines. As soon as the timer closes the primary circuit and the current begins to flow, the core of the relay attracts its armature, thereby breaking the primary circuit at the contact points, and causing a spark to be produced in the secondary circuit. The opening of the circuit at the contact points causes the current in the coarse winding of the relay to cease, but the current in the fine winding is not affected, and the magnetic effect of this current keeps the armature in position down on the core and prevents its vibration. It is held in this position until contact is broken in the timer, when the current in the fine wire winding ceases and the armature returns to its position of rest.

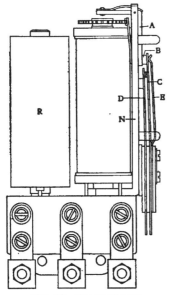

FIG. 99—DELCO RELAY.

The only adjustment on the Delco relay is at F. By turning the notched wheel the gap between the armature and the end of the magnet core is varied, and as this latter acts as a stop the opening at the contact points is also varied. The adjustment is locked by a latch spring, which engages in the notches in the adjusting wheel.

High Speed Interrupter—Recently the normal working speed of motors has in some cases been increased to an extraordinary extent, and this has made the demand upon ignition systems more severe than ever. It is generally believed that a magnetic vibrator system of ordinary construction is not sufficiently rapid in action to meet the requirements of a true high-speed engine. Even mechanical interrupters of the ordinary design have proved defective at very high speeds, as is shown by several patents recently issued on improved high-speed interrupters. In a system recently patented by C. F. Kettering, inventor of the Delco system, two relays are used in connection with a six-cylinder motor, and an interrupter cam having three lobes and two oppositely located sets of contact points, the stationary point of each set being connected to one of the relays. The arrangement is shown in diagram in Fig. 101. It will readily be seen that the circuit

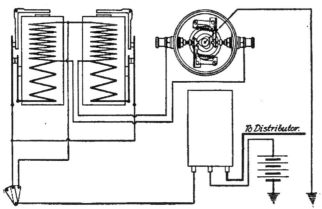

FIG. 100—DELCO DOUBLE INTERRUPTER SYSTEM FOR HIGH SPEED WORK.

is closed through the two relays alternately, and this gives each magnetic interrupter more time to complete its function of opening and closing the circuit.

Other Types of Closed Circuit Interrupters—It was the need for reliable operation at high speed that led to the development of the modern "closed circuit" interrupter. To insure quick action, the moving part of the interrupter must be made to have the least possible inertia. Fig 102 shows two such interrupters, the new Delco and the Atwater Kent. Both of these are operated by a cam on the interrupter shaft. The main difference is that in the Delco the contact point is carried at the end of the lever and the cam follower is between the contact point and the pivot, whereas in the Atwater Kent the arrangement is the reverse. Besides,

MODERN BATTERY SYSTEMS 129

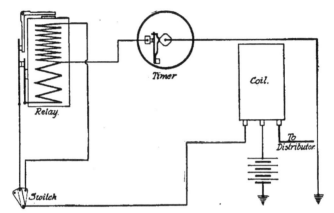

Fig. 101—Starting and Running Connections of Delco Relay System.

the Delco lever is pivoted, whereas the Atwater Kent lever is spring-mounted.

A somewhat unusual system has been introduced by the Phillips-Brinton Co. under the name of the Philbrin high-frequency system. It comprises a magnetic vibrator which vibrates continuously as long as the switch is closed, and a high-tension distributer which also performs the function of a timer. The rotating distributer arm ends in a sector which extends from one stationary contact to the next, so

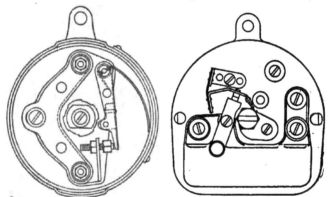

Fig. 102—Two Recent Types of Closed Circuit Interrupter (Delco and Atwater Kent).

the spark, as soon as it stops in one cylinder, begins in the next.

The condenser is now generally placed either in or on the interrupter, instead of with the coil. As the condenser must be connected in shunt across the interrupter this obviates the need for any outside wiring for this purpose. A neat arrangement of the condenser is found in the Philbrin interrupter illustrated in Fig. 103. It is enclosed in a metal case and is moisture proof and protected against mechanical injury.

Non-Vibrator Coils—Coils for mechanical interrupter battery systems have been reduced to a very simple and compact form. They are enclosed in cylindrical cases of Bakelite or

FIG. 103—PHILBRIN INTERRUPTER, SHOWING MOUNTING OF CONDENSER.

drawn steel and are about 2 inches in diameter by 6 inches long. They are generally mounted on the engine close to the combined interrupter and distributer so that both the primary and secondary connections are very short.

Current Consumption—As already stated, the open circuit system of battery ignition was developed in response to a demand for high current economy, and in general the open circuit systems require considerably less current than the closed circuit systems. Their characteristics as to change of current consumption with change in engine speed are also entirely different, as may be seen by reference to Fig. 104. Since in the open circuit system the length of contact is independent of the engine speed, the electrical energy absorbed by the primary per

MODERN BATTERY SYSTEMS 131

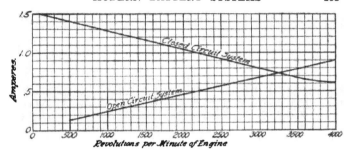

Fig. 104—Current Consumption of Four Cylinder Open and Closed Circuit Ignition Systems at Varying Engine Speeds.

contact is constant, and the current increases in direct proportion to the number of contacts per unit of time; in other words, in proportion to the engine speed. On the other hand, the closed circuit system shows its greatest consumption of current when the engine is at a standstill and the switch is closed. A steady current depending upon the resistance of the coil primary winding then flows, of the maximum value the current can reach in the primary circuit under the impressed electromotive force. When the engine runs the circuit is interrupted and the current assumes an intermittent character, and the oftener it is interrupted in a given time the lower the average value of the current wave. With the particular system from which the test results were obtained the current flowing through the coil with the engine at a standstill amounted to 2.9 amperes. As soon as the engine begins to run this drops to 1.5 amperes, and it decreases further with an increase in engine speed until at 3000 r.p.m. of the engine it is down to 0.8 ampere. With the open circuit system the current consumption is only 0.25 ampere at 1000 r.p.m. and 0.68 ampere at 3000 r.p.m. Both curves are for four cylinder systems operating on three-cell storage batteries.

CHAPTER XIII

Magneto and Coil Ignition

Of the different magneto systems the ones most closely related to the battery and coil system are those comprising a magneto generator furnishing low tension current impulses which are transformed into high tension impulses in a coil or coils separate from the magneto. There are various possible combinations. Thus, we may combine a plain generator of current impulses with a set of vibrator coils equal in number to the number of cylinders to be fired; we may combine this same generator with a single vibrator coil and a high tension distributer which would preferably be incorporated with the magneto, or we may combine a magneto provided with a mechanical interrupter and a high tension distributer with a plain or non-vibrator coil. In the first two cases a timer would also be required and this might either be incorporated with the magneto or be separate from it. Only the first and the last of the three combinations enumerated are used in practice. The first, that in which a plain generator of current impulses is combined with a set of vibrator coils, is exemplified in the Ford ignition system.

Ford Flywheel Magneto—The Ford magneto, which is built with the engine flywheel, is of unusual construction. It consists of 16 V-shaped magnets which are clamped to the forward side of the flywheel, like poles of adjacent magnets being placed together, so that the set really forms a single 16-pole magnet. To a pressed steel disc located directly in front of the flywheel are secured 16 coils provided with soft iron cores, these being equally spaced in a circle of such diameter that the magnet poles sweep past the ends of the cores as the magnet rotates with the flywheel, the air gap between the magnet poles and the armature cores being only 1/32 inch. The coils are wound with copper ribbon and are connected in series, alternate coils being wound in opposite directions. The magnetic circuit formed by the magnets, the cores of the coils and the steel disc to which the coils are fastened, is illustrated in the diagram Fig. 107. In

MAGNETO AND COIL IGNITION 133

this magneto the usual order of things is reversed in that the magnets rotate and the armature is stationary. It will readily be seen that when a certain coil is opposite a north pole of the magnet, the magnetic flux passes through it in one direction, and when the same coil comes opposite the next south pole, the magnetic flux passes through it in the opposite direction. Hence in passing from one pole to another, the magnetic flux through the coil is completely reversed, and it is this reversal of the flux that induces an electromotive force in

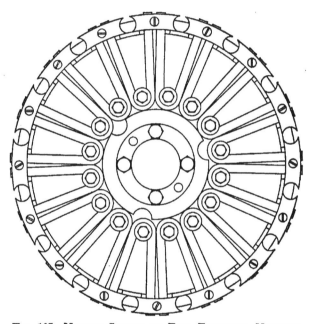

FIG. 105—MAGNET SYSTEM OF FORD FLYWHEEL MAGNETO.

the winding of the coil. If all the coils were wound in the same direction, the electromotive forces in those coils approaching, say, a north pole, would be opposite in direction to the electromotive forces in the coils approaching a south pole. By winding adjacent coils in opposite directions, the electromotive forces induced in all of the coils are alike in direction at all times, hence the electromotive forces of the individual coils add together. Of course, when a particular coil approaches a north pole the electromotive force induced in it

134 MAGNETO AND COIL IGNITION

is opposite in direction from that induced in it when the same coil approaches a south pole, and the magneto, therefore, generates an alternating current.

In connection with the magneto there is used a roller contact timer of the type already described. This timer has four terminal posts from which wires are run to the ground terminals of the four vibrator coils respectively. One end of the armature winding of the magneto is grounded and

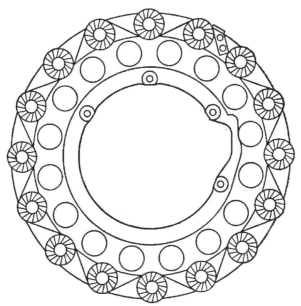

FIG. 106—STATIONARY ARMATURE OF FORD MAGNETO.

the other connects through a switch to the primary terminal of the vibrator coils. A diagram of connections of the Ford system is shown in Fig. 108.

Characteristics of Ford System—There is one important difference between the Ford system and an ordinary coil and battery system, namely, that the Ford magneto furnishes an alternating current which alternates 16 times per revolution of the crankshaft. Each time the direction of the current changes there is, of course, a momentary cessation of current. It is obvious that in the positions of the crankshaft and field magnet for which there is no current in the arma-

MAGNETO AND COIL IGNITION 135

FIG. 107—MAGNETIC CIRCUIT OF FORD MAGNETO.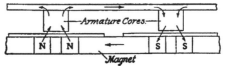

ture or when the current is only very small, it is impossible to produce a spark in the cylinders. Suppose that when the timer completes the primary circuit the magneto is at or very near the position of zero electromotive force. Then the coil vibrator cannot act at once, as the current sent through the coil by this small electromotive force is too weak to set it in vibration. But as soon as the current has attained the minimum value required to operate the vibrator a spark is produced. The result of this is that as the timing lever is moved over its quadrant the spark is not advanced uniformly with the lever motion. However, the angular motion of the magnet for which there is insufficient current to produce a spark is so small that the condition explained does not affect the practical operation of the system.

The Ford magneto is of comparatively large size and produces a spark at a low speed of revolution. It furnishes the spark for starting the motor as well as for regular operation.

This magneto has no brushes and, in fact, absolutely no wearing parts, hence it is subjected to few troubles. One thing that occasionally does occur is that the magnets lose their strength. A current sent through the armature coils

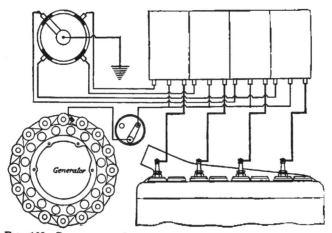

FIG. 108—DIAGRAM OF CONNECTIONS OF FORD IGNITION SYSTEM

136   MAGNETO AND COIL IGNITION

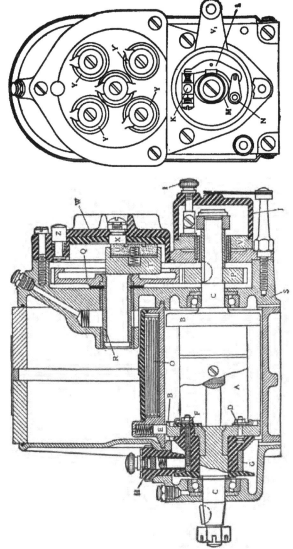

Figs. 109 and 110—Longitudinal Section and End Elevation of Remy Model P Magneto.

from an outside source will have a tendency to either strengthen or weaken the magnets, according to the relative position of armature and magnet and the direction of the current. As the armature is always likely to be in such a position that the current sent through it from an outside source would weaken the magnets, no battery should ever be connected to the armature.

The voltage of the Ford magneto increases with the engine speed, being 10 at 300 r.p.m., 16.5 at 600 r.p.m., 24 at 1000 r.p.m., and 30 at 1500 r.p.m.

Magneto and Non-vibrator Coil System—A typical magneto of the class designed for use with a separate coil is the Remy Model P, of which a sectional view and an end elevation are shown in Figs. 109-110. Referring to this cut, the armature core A is of the H or shuttle wound type and is built up of annealed sheet iron discs with hard bronze heads BB which are cast on to the shafts CC. After the core has been insulated it is wound with cotton covered, enameled magnet wire, which is impregnated with an insulating compound to render it moisture proof and capable of withstanding high temperatures. The armature is carried in two ball bearings, as shown. One end of the armature winding is connected to the rear head at D, and effective grounding of this head to the frame of the magneto is assured by the spring pressed carbon brush E, carried in a brush holder formed integral with the armature cover plate. The other end of the armature winding is connected through an insulated stud F to the collector ring G, from which the current is taken off by a carbon brush in the brush holder H. From this brush holder a low tension cable is run to the induction coil.

Interrupter—Another wire runs from the induction coil to the terminal I on the "breaker box" at the other end of the magneto. This terminal, which extends through the interrupter cover J of insulating material, forms an integral part of the contact screw K, which carries one of the contact points of the interrupter. The other contact point is carried on the free end of lever L, pivoted at its lower end and provided at its middle with a fiber contact block bearing against the cam M carried on the end of the armature shaft.

All parts of the interrupter, viz., the contact screw K, the interrupter lever L and its stud N are supported on a metal plate V forming the interrupter base. This plate is supported on a lateral projection from a disc secured to the forward end plate S of the magneto and is provided with a radial arm V, which is connected by a linkage to the spark lever on the steering post of the car. By means of this arm the interrupter base can be moved angularly around its axis and the spark thus caused to occur earlier or later in the cycle of the engine.

Interrupter lever L is grounded to the frame of the magneto through the stud N. Condenser O is located in the armature cover plate. It has one terminal connected to the stationary contact screw K and the other terminal grounded, so that it is connected directly across the interrupter.

High Tension Distributer—Upon the armature shaft between the armature and the interrupter cam is carried a steel spur pinion P which meshes with a bronze spur gear Q of twice its number of teeth, secured to a short shaft R supported in a plain bearing in the bronze end plate S of the magneto. Rigidly mounted on the bronze distributer gear Q is a brush holder T which carries a carbon brush U in a recess formed in its face. This brush is pressed by a coiled spring against the inner surface of the distributer plate or cover W of Bakelite in which are embedded a central contact block X and four or six (according to the number of cylinders to be sparked) contact blocks YY arranged in a circle and at equal angular distances. Each contact block carries a binding screw on the outside of the distributer cover, protected by a ridge of insulating material all around its circumference except at or near the top where the high tension cables lead out. As the distributer brush revolves it makes contact between the central contact block X and all of the contact blocks Y in succession. Contact blocks Y are connected by high tension cables to the spark plugs.

It will be seen from Fig. 110 that the interrupter cam M has two lobes. The magneto, which is designed for use on a four-cylinder, four-cycle engine, is geared to run at crank shaft speed. It produces two electromotive impulses per revolution and the cam will open the circuit each time the electromotive impulse is at or near its maximum.

Spark Timing and Timing Range—Ordinarily the interrupter contacts are closed, being held together by a spring not shown in the drawings, and as soon as an electromotive impulse in the armature begins, a current will begin to flow through the armature winding and the primary winding of the coil. At about the time when the receding edge of the armature pole passes the farther edge of the field pole, the circuit is broken at the interrupter contacts by cam M moving lever L outwardly, and the current through the primary winding of the coil ceases. This induces a high tension impulse in the secondary winding of the coil and causes a spark to pass at one of the spark plugs. The exact time at which this occurs depends, of course, upon the setting of the interrupter base by means of the spark lever. The most effective spark is produced if the circuit is broken at the moment when the current in the circuit has just attained its maximum value. It was explained in a previous chapter and illustrated in Fig. 59 that the current impulses in the armature winding are of very short duration, the current increasing and decreasing very rapidly. The spark control linkage is usually so arranged that when the spark lever is set in the "maximum advance" position the circuit is interrupted when the primary current has almost reached its maximum value. If now the spark lever is moved to the "late spark" position the circuit will be interrupted some time after the current has passed its maximum value and in consequence the intensity of the spark for the same armature speed will be reduced. The more the spark is retarded the more it is reduced in intensity. This consideration limits the possible timing range. In the magneto here illustrated the timing range is 35 degrees.

Distributer Action—Since the distributer gear Q has twice as many teeth as armature pinion P, the distributer makes one revolution for every two of the armature or of the crankshaft. In other words, the distributer brush makes one revolution during the time of one engine cycle and makes contact with each of the four distributer blocks Y once per cycle. The gears are so enmeshed that the distributer brush U is on one of the contact blocks Y whenever the primary circuit is interrupted. The exact moment of the break, of course, varies with the setting of the spark lever, but the distributer brush U is made so wide that without any angular displacement of the distributer cover the contact blocks will be opposite it when the spark occurs throughout the whole timing range.

A feature of this particular magneto is a timing button Z fitted into the distributer cover. This button is pressed outwardly by a coiled spring underneath it. If the button is pressed inward by the thumb against the pressure of the spring and the magneto armature is then turned, the plunger of the button will drop into a recess of the distributer gear. Then the engine crankshaft must be turned until the piston in cylinder No. 1 is in the top dead center position at the beginning of the firing stroke, and while the engine crankshaft and magneto armature shaft are in these respective positions the magneto driving gears must be meshed and the

140 MAGNETO AND COIL IGNITION

magneto gear secured on the tapered portion of the armature shaft by Woodruff key and nut.

Connections of Dual System—An ignition system of this kind is generally arranged for dual operation. That is, a battery of dry cells or a storage battery is carried, and the current from this battery may be used instead of the magneto current, the battery current being interrupted, transformed, distributed and used in the form of ignition sparks by the same devices as the magneto current. Hence only the source of current is duplicated in this dual system.

A diagram of connections for this system is shown in Fig. 111. It will be seen that the coil is provided with a switch

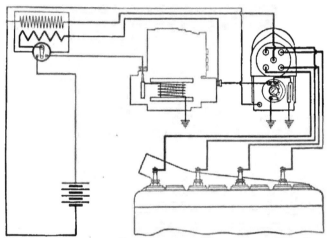

FIG. 111.—DIAGRAM OF CONNECTIONS OF MAGNETO AND COIL, DUAL SYSTEM.

having three positions, viz., "Battery," "Off," and "Magneto." When the switch handle is in the "battery" position the current flows from the battery through the switch, the primary of the coil and the interrupter into ground and back through the grounding wire of the coil to the battery. The secondary current, originating in the secondary winding of the coil, flows through the high tension cable to the central terminal of the distributer, through the distributer and spark plug cable to one of the spark plugs, thence into ground and through the grounding wire back to the secondary winding of the coil.

If the switch lever is turned to the "magneto" position,

MAGNETO AND COIL IGNITION 141

the current from the magneto armature flows through the switch, the primary winding of the coil and through the interrupter into ground, whence it returns to the armature winding, one end of which is grounded. The flow of the secondary current is the same when the primary current is furnished by the magneto armature as when it is furnished by the battery. The chief object in carrying the battery is to facilitate starting of the motor. With the battery current the intensity of the spark is independent of the engine speed, but with the magneto current the intensity of the spark falls off with the speed. Of course, the battery will also serve as a reserve source of current in case the magneto should fail to generate for any reason, while coil, interrupter and distributer are performing their functions properly, but this possibility is comparatively remote.

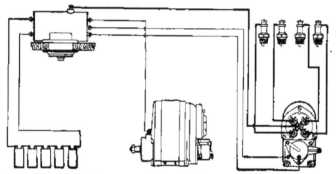

FIG. 112—WIRING DIAGRAM OF REMY MAGNETO AND COIL SYSTEM.

Wiring Diagram—Fig. 112 shows a so-called wiring diagram for this system which is of help in installing it and in re-establishing connections if they should have been unmade for any reason. As a help in tracing out connections, each of the wires has an insulating covering of different color, a practice now widely followed in automobile electric work. It will be seen that the switch is formed integral with the coil, the former being located on the rear side of the dash and the latter on the forward side. It is now the general practice to carry single coils on the forward side of the dash, under the engine hood, where they are better protected from moisture and mechanical injury and where they are out of sight. In connecting the battery, it does not make any difference to which of the two coil binding posts the positive terminal of the battery is connected, unless the battery used as a reserve ignition current source is also used for other purposes, on a circuit having a ground return. In that case it is necessary that the same side of the battery which is grounded by the ground return of this circuit be also the one grounded by the coil ground wire, as in the alternate case both terminals of the battery would be grounded and the battery would be short-circuited.

CHAPTER XIV

The High Tension Magneto

A high tension magneto is one which furnishes a jump spark without the intermediary of a separate coil. The armature of such a magneto, in addition to serving the purposes of an ordinary armature, also acts as a coil or step-up transformer. In fact, a high tension magneto with its interrupter and distributer forms a complete ignition system, the only outside parts being the spark plugs and the switch.

History of the Device—This type of magneto was invented by Paul Winant, of Deutz, Germany, in 1888. It seems, however, that the technical difficulties involved in the construction of such a magneto were too great for the state of the electrical art at that early period, for, as far as is known, no extensive use was made of high tension magnetos for engine ignition until Robert Bosch took up their manufacture about 1903. Bosch had previously for many years manufactured magnetos for low tension make and break ignition.

Bosch Magneto—A sectional view of the latest type of Bosch high tension magneto (Model ZU4) is shown in Fig. 113. The armature core is built up in the same manner as for a low tension magneto, as already described. After the insulation has been put on the core the primary winding A is wound on, the beginning of this winding being grounded to the core. The secondary winding B is wound right over the primary, the beginning of the secondary being connected to the end of the primary.

On a high tension magneto the condenser is usually incorporated in the armature, as this simplifies the connections, and in order to reduce the bulk of the condenser, mica is used for the dielectric. In this case the condenser C is located inside the cup-shaped armature head b_1. A lead d is brought out from the end of the primary winding to one terminal of the condenser, and the other terminal of the condenser is grounded to the armature core. The grounding of the primary winding and of one side of the condenser is indicated at a.

144   THE HIGH TENSION MAGNETO

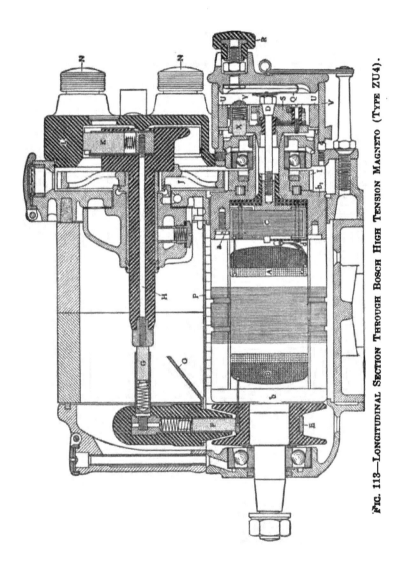

Fig. 113—Longitudinal Section Through Bosch High Tension Magneto (Type ZU4).

The insulated side of the condenser is also connected to the interrupter fastening screw D. As in other types of magneto with shuttle wound armature, the armature core is provided with two heads, b b_1, the one at the driving end being formed integral with a solid shaft with a tapered portion to which a driving coupling may be secured, and the one on the interrupter end with a hollow shaft through which extends the insulated interrupter fastening screw D. Both of these shafts are supported in radial ball bearings.

The end of the secondary winding B is connected to the collector ring E embedded in an insulating ring, with high flanges on both sides, on the driving shaft of the armature. Upon this collector ring bears a carbon brush F carried in a brush holder secured by screws to the end plate of the magneto. The brush holder consists of a block of insulating material molded to metal fittings, of which one forms the holder for the brush F, already referred to, and the other the holder for a brush G bearing against the head of the distributer pin H.

High Tension Distributer—Mounted on the armature head b_1 is a spur pinion I which meshes with a gear J on the distributer shaft, located directly above the armature, the gear having twice the number of teeth as the pinion and therefore rotating at only one-half its speed.

The distributer shaft is made hollow and through it extends the distributer pin H, insulated from it by hard rubber. Pin H screws into the metal part of the distributer brush holder which contains the carbon brush K. The latter is pressed by a coiled spring against the internal cylindrical surface of the distributer disc L of molded insulating composition. Embedded in this disc are four contact segments M which are in electrical connection with binding posts N on the outside of the distributer disc, from which connection is made to the spark plugs.

As the distributer brush K rotates it establishes electrical connection with the four contact sectors M in succession. It will be noted in Fig. 115 that each contact sector covers a considerable arc, which insures that brush K will be in contact with a particular sector whether the spark is timed to occur early or late.

Safety Spark Gap—In order to prevent injury to the armature windings from excessive voltages, as might occur if one or more of the high tension cables leading to the spark plugs should become detached and the magneto be operated at very high speed, a device known as a safety spark gap is provided. This consists merely of an inclined strip of metal O secured

146   THE HIGH TENSION MAGNETO

to the armature cover plate P and approaching to within ⅜ inch of the holder for brush G. Since one end of the secondary winding is grounded and the other end is directly connected to brush G, it is obvious that whatever electromotive force is induced in the secondary winding tends to cause a spark to jump between terminal O and brush holder G. Ordinarily, when all of the high tension terminals leading to

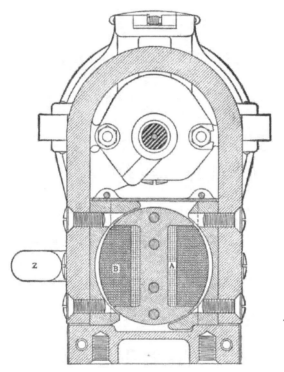

Fig. 114—Cross Section Through Bosch High Tension Magneto.

the spark plugs are connected and the spark points of the plugs are properly adjusted, the resistance to the spark is less at the gap of the spark plugs than at the safety spark gap, and the spark will pass at the plug. But if the spark for any reason cannot jump across the gap at the plug, it will jump at the safety spark gap and thus protect the armature against excessively high electromotive force.

THE HIGH TENSION MAGNETO 147

Interrupter—Against the end of the interrupter fastening screw D bears a conducting strip Q, which establishes electrical connection between the fastening screw and the terminal post R on the outside of the interrupter box. The fastening screw D has a conical seating in, and therefore is electrically connected with, the support S of the stationary

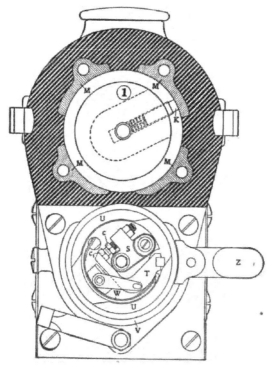

Fig. 115—End View of Bosch Magneto with Distributer Shown in Section.

contact c of the interrupter. The movable contact c_1 of the interrupter is carried on one arm of the bell crank T, whose other arm is provided with a fiber block adapted to engage with cam lobes U U on the inner surface of the interrupter box V. The two cam lobes are located opposite each other, and when the fiber block rides on them the contact points c c_1 are separated, while ordinarily they are held together by

148 THE HIGH TENSION MAGNETO

the bent spring plate W. Effective grounding of the interrupter lever T is insured by a brush X carried in a brush holder formed integral with the interrupter base Y. Variation of the moment of breaking the circuit is effected by means of a lever arm Z secured to the interrupter shell V. Shell V and its cover plate are held in position by a flat spring supported on a stud projecting from the forward end plate of the magneto, and they therefore can be very quickly removed and replaced.

A diagram of connections of the Bosch high tension magneto is shown in Fig. 116. All of the principal parts of the machine

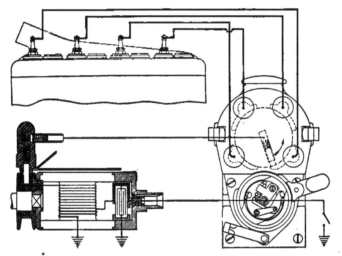

FIG. 116—DIAGRAM OF CONNECTIONS FOR BOSCH HIGH TENSION MAGNETO.

are shown in their proper place. The circuits can be very easily traced out and it is unnecessary to explain the paths followed by the primary and secondary currents at length. From the terminal post on the interrupter box of the magneto a wire is led to the magneto switch, which may be located on the dashboard or the steering post of the car. The other terminal of the switch is simply connected to ground, and when the switch is closed the primary winding of the armature is permanently short-circuited, the interrupter no longer having the effect of opening it. This is clearly shown in Fig. 117. The inductive effect of the primary winding on the secondary winding then is reduced to such an extent that no spark is produced.

Details of Construction—A few details of construction may here be pointed out. The magneto is completely enclosed, the end plates where they join the magnets being provided with grooves containing packing material, which insures a dust and water proof joint. Over the bearings there are oil holes with hinged caps. In the lower parts of the housings there are drain holes through which any excess oil may drain off. The plain bearing of the distributer shaft is lubricated by the so-called wick system. The distributer disc is held in position by two spring clips and can be quickly removed for inspection, cleaning, etc. It is provided with a mica window through which the rotation of the distributer brush may be observed. This window serves for timing the magneto. On the revolving part of the distributer is painted a figure 1 in a circle. In setting the magneto the

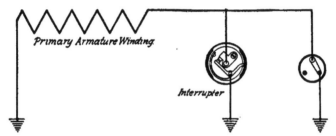

FIG. 117—DIAGRAM OF PRIMARY CIRCUIT.

engine crankshaft is first turned so piston in cylinder No. 1 is in the dead center position at the end of the compression stroke, and is then turned back from 22 to 34 degrees (depending on the particular engine). The armature of the magneto is then turned until the figure 1 shows in the mica window. Then the oil well cover over the distributer gear is lifted, when the distributer gear J can be seen through a sight hole. The armature is then turned farther until a marked tooth of the distributer gear registers with marks on the sides of the sight hole. With the armature in this position the magneto drive is connected, great care being taken not to move the armature during the operation.

In many magnetos only the armature space is enclosed and it is then necessary to use a special design of safety spark gap, as, if a spark should occur in the open space between the magnets, any gasoline vapor that might be present

FIG. 118 — SAFETY SPARK GAP.

would probably be ignited and an explosion would occur. The safety gap used in such cases takes the form shown in Fig. 118. A cylindrical chamber is cast integral with the armature cover plate. One terminal, the grounded one, is secured centrally into the bottom of the cylinder and the other into the insulated cover plate. This latter is connected to the high tension brush holder. Through the wall of the cylindrical chamber are drilled six holes which are covered with fine mesh wire gauze. This permits any sparks in the safety device to be seen, but no flame to be propagated through the wire gauze.

Principle of Operation—When the armature of the high tension magneto rotates, electromotive forces are induced in both the primary and secondary windings. Since the primary winding is short-circuited upon itself by the interrupter, and the resistance of the circuit therefore is very low, a heavy current will be established in that winding. On the other hand, the electromotive force induced in the secondary winding, though many times greater than the primary electromotive force, is not sufficient to cause a spark to jump at the spark plug to which the insulated terminal of that winding may be connected at the moment by the high tension distributer. Hence the secondary circuit is open and no current flows in it. When the current in the primary winding has about attained its maximum value—somewhat earlier or later according to the setting of the spark lever—the primary circuit is interrupted, and the cessation of current in the primary winding produces a strong inductive effect on the secondary winding. The electromotive force induced in the secondary winding by the cessation of the primary current is in the same direction as that induced in it by the revolution of the armature, hence the two inductive effects add together and produce an electromotive force of sufficient strength to cause a spark to jump across the gap of the spark plug.

There is one fundamental difference between the spark produced by a true high tension magneto and that produced by a coil system. When a separate coil is used the energy of the spark is limited by the magnetic energy stored up in the core of the coil, since there can be no further supply of energy from the source of current after the primary circuit is broken. In the case of a high tension magneto, however, after the primary circuit has been opened and the arc at the spark plug established, the rotation of the armature

THE HIGH TENSION MAGNETO 151

continues to generate current in the secondary winding, and thus tends to prolong the discharge at the spark plug terminal; in other words, to increase the duration of the spark and consequently to increase the quantity of electrical energy

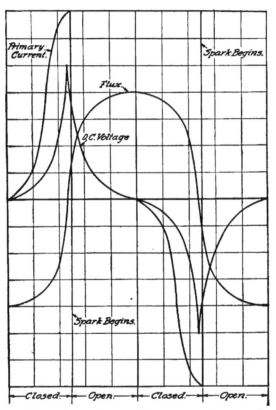

Fig. 119—Flux, Open Circuit Voltage and Primary Current Curves of High Tension Magneto.

involved in it. On account of the greater duration of the spark from a high tension magneto the latter is sometimes called an arc flame magneto.

Forms of Flux, Voltage and Current Waves—A voltage curve of a magneto generator was shown in Fig. 59. In Fig. 119 is given a similar curve together with a curve showing the flux through the armature core and the current flowing in the primary winding of the armature. The outstanding fact shown by this diagram is the instant cessation of the primary current, which corresponds in time to the beginning of the spark at the plug. This diagram is based on one given in a paper on "The High Tension Magneto" read by A. P. Young before the Aeronautical Society of Great Britain in 1917.

CHAPTER XV

Special Types of Magnetos—Care and Adjustment

We have so far described two distinct classes of ignition magnetos, in one of which a wire-wound armature rotates inside a stationary magnet frame, and in the other of which a magnet frame rotates past a stationary armature. There is a third class, known as inductor magnetos, in which both the magnet frame and the armature winding are stationary and only a part of the magnetic circuit, called the rotor, rotates. The magnet frame of such a magneto may be similar to that of the ordinary magneto with stationary magnet frame, but it may also be quite different.

Rotary Sector Type—There is considerable variation in the form of the rotor in inductor magnetos. In the earliest form of this magneto for ignition purposes (Bosch), which is now obsolete, the magnet frame and armature had the same general form as in a modern rotating armature magneto, but the armature was of considerably smaller diameter than the bore of the magnet poles, and the space between armature and magnet poles was occupied by two sectors of magnetic material, secured to end plates with shafts and adapted to rotate around their axis, the armature being fixed in the position which in a rotary armature magneto corresponds to maximum induction (that is, armature poles midway between magnet poles). If it were not for the rotating sectors, no magnetic flux would pass through the armature winding. In Fig. 120 the sectors are shown in two successive positions 90 degrees apart. It will be seen that in the first position the magnetic lines of force pass through the armature core from top to bottom and in the second they pass through in the opposite direction. The two positions represented are positions of maximum flux. Hence, the direction of the flux through the armature winding is reversed every 90 degrees, or four times per revolution of the sectors, as compared with twice per revolution of the armature in a revolving armature machine.

154 SPECIAL TYPES OF MAGNETOS

An inductor magneto of this particular type therefore has to revolve only half as fast as a revolving armature magneto under otherwise similar conditions. It is hardly necessary to further explain the inductive action, as according to the laws of electromagnetic induction, stated in a previous chapter, variation of the magnetic flux through a coil of wire always induces an electromotive force therein.

Remy Low Tension Inductor Magneto—Another type of inductor magneto, which has been very extensively used in this country, the Remy, is represented by the diagrams Fig. 121. This magneto has a rotor composed of a central core and two pole pieces, the latter being arranged opposite to each other or 180 degrees apart. Over the central part of the core is

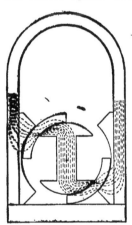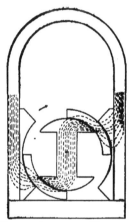

FIG. 120—DIAGRAM OF ROTATING SECTOR INDUCTOR MAGNETO.

located the armature winding in the form of an annular coil of sufficient internal diameter to clear the core. The parts of the rotor are secured to a shaft mounted in ball bearings as usual. The armature coil fits snugly to the magnet poles and is supported by them. Referring to Fig. 121, in subfigure 1 the front pole of the rotor is adjacent to pole N of the magnet, and the rear pole of the rotor is adjacent to pole S of the magnet. The magnetic circuit is completed through the shaft and the central core of the rotor. The magnetic flux passes from magnet pole N to the front rotor pole, through the central core and shaft to the rear rotor pole and thence to magnet pole S. Thus the magnetic flux passes through the armature coil from front to back. In subfigure

SPECIAL TYPES OF MAGNETOS

2 the rotor is so located that each of its poles forms a path between poles N and S of the magnet, and the magnetic flux passes from one pole to the other without passing through the armature coil. In subfigure 3 the conditions are similar to those in subfigure A, but the front rotor pole is now adjacent to magnet pole S and the rear rotor pole to magnet pole N, so that the magnetic flux passes through the armature coil from back to front, or in the opposite direction as in subfigure 1. In subfigure 4 conditions are similar to those in subfigure 2, except that the front rotor pole is down and

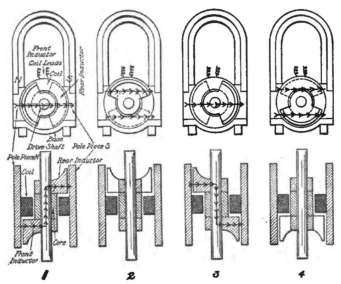

FIG. 121—DIAGRAMS OF REMY INDUCTOR MAGNETO.

the rear one up. In this position again no magnetic flux passes through the armature coil. It will thus be seen that as the rotor revolves the magnetic flux through the coil varies continually, and as a result an electromotive force is induced in the armature coil.

As in the case of the rotating armature magneto, the electromotive force in the armature winding attains its maximum value when there is no magnetic flux passing through the armature winding, as in positions 2 and 4. On the other hand, the current induced in the winding reverses its direction when the flux through it is at its maximum, as in positions

SPECIAL TYPES OF MAGNETOS

1 and 3. The induced electromotive force in this type of inductor magneto therefore is reversed twice for every revolution of the rotor, the same as in a rotating armature magneto.

Since the armature coil is stationary, no collector ring and brush are required to take the current from the armature and the inductor magneto is generally the simplest type possible.

K-W High Tension Inductor Magneto—Inductor magnetos

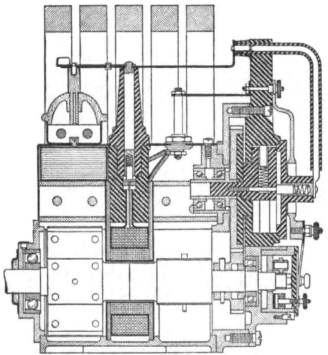

FIG. 122—LONGITUDINAL SECTION OF K-W HIGH TENSION INDUCTOR MAGNETO.

may be either of the low tension or the high tension type. If low tension, they have only a single armature coil, as above described, while if high tension, they have both a primary and a secondary armature winding, the latter being usually wound over the former.

Fig. 122 is a longitudinal section of the K-W high tension inductor type magneto and Fig. 123 a sketch of its magnetic

SPECIAL TYPES OF MAGNETOS

circuit. The two magnet poles of this magneto, instead of being located on opposite sides of the rotor or 180 degrees apart, are located at quarters, or 90 degrees apart, and the rotor is built up of thin laminations of sheet iron which are riveted together, bored out, turned off and secured to their shaft by taper pins. There are two of these blocks of laminations and they are secured to the shaft at right angles to each other. The rotor, therefore, has four poles, and it can easily be shown that in this magneto four impulses are induced per revolution. Being driven at crankshaft speed, the magneto can be used for firing any motor from a single cylinder to a four-cylinder two-cycle and from a two-cylinder to an eight-cylinder four-cycle. The cam used for operating the interrupter, carried at the end of the rotor shaft, is a cylinder with one or more flats on it—one flat for a two-cylinder, two flats for a four-cylinder, and four flats for an eight-cylinder motor (all four-cycle). In the case of the two-cylinder motor only one-quarter of the impulses and in case of the four-cylinder only one-half the impulses are utilized. During the other electromotive impulses the interrupter does not operate, and no spark is produced.

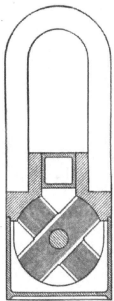

FIG. 123—DIAGRAM OF MAGNETIC CIRCUIT OF K-W INDUCTOR MAGNETO.

Leads from the primary winding extend downwardly and pass out through the base to the interrupter, and a lead from the secondary winding extends upwardly and connects to the high tension distributer and the safety spark gap. All of the accessory devices, such as interrupter, distributer and safety spark gap, operate on the same principle as other devices of the same class already described, differing from them only in details. This particular magneto is claimed to generate a sine wave of current, which permits of a comparatively large timing range.

Dixie Magneto—A peculiarity of the Dixie high tension magneto, manufactured by the Splitdorf Electrical Company, is that the permanent magnets of U form straddle the length of the revolving part or rotor, so that the axis of the rotor

158 SPECIAL TYPES OF MAGNETOS

lies in the central plane of the field magnets, instead of being perpendicular thereto. Both windings—primary and secondary—are stationary, and current is induced in them by the rotation of the rotor, which consists of a central part of non-magnetic material with pole extensions of magnetic material on both ends, the whole being secured to short shafts carried in bearings supported by the field magnets. In addition to the stationary field frame, consisting of two U-shaped magnets, and the rotary element or rotor, there is in this magneto a third member of the magnetic circuit, consisting of the core on which the coils are wound, and two pole shoes secured thereto. This core and the pole shoes are built up of laminations of soft sheet iron. The two polar extensions come quite close to the inside surfaces of the two poles of the field magnets respectively. The path of the flux through the magnetic circuit is clearly shown in the diagram Fig. 124, from which

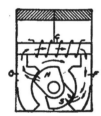 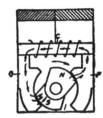

Fig. 124—Magnetic Circuit Diagram of Dixie Magneto.

it may be seen that the flux through the coils is completely reversed in direction twice during every revolution of the rotor. In Fig. 125 is shown a section through the rotor axis of the Dixie magneto. For eight and twelve-cylinder engines the field poles are placed at quarters and double polar extensions, set crosswise of each other, are provided, the principle employed being the same as that embodied in the K-W magneto already described.

Teagle Inductor Magneto—Another unusual magneto is the Teagle, manufactured in Cleveland, Ohio. It employs straight bar magnets which are carefully ground and clamped to top and bottom yokes. The bottom yoke forms a pole piece extending nearly half way around the rotor tunnel, while the top yoke has two poles, one carrying the windings or coils, and the other serving as a magnetic by-pass. The magnetic circuit is shown diagrammatically in the two accompanying sketches Fig. 126. In one the rotor pole is shown opposite the pole of the top yoke which carries the windings or coils, and with the rotor in this position, the maximum flux passes through the coils. In the other sketch the rotor is shown opposite the end of the magnetic by-pass, and in this position all of the magnetic flux passes through the by-pass and none through the coils. As the rotor has two poles, the flux through the coils passes through a maximum and a minimum twice during every revolution of the rotor.

Magnetos for V Type Engines—In most types of engines explosions occur at regular intervals, and as the ordinary magneto generates current impulses separated by equal time intervals, by selecting a suitable gear ratio between engine

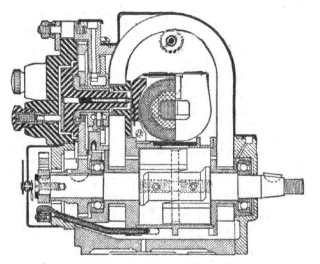

Fig. 125—Section Through Dixie Magneto.

crankshaft and magneto armature shaft, the ignition requirements of both two and four-cycle engines of any number of cylinders can easily be met. It is different, however, with the so-called V type engine often used for motorcycles, in which two cylinders are set at an angle of 90, 60 or 45 degrees. If the angle between the two cylinders of a V motor is designated by a, then the explosions in the two cylinders follow each other at irregular intervals corresponding to crankshaft motions of $360 + a$ degrees and $360 - a$ degrees, respectively. One method of solving the problem of producing sparks at unequal intervals corresponding to the intervals between explosions is to use two separate armatures arranged end to end in

SPECIAL TYPES OF MAGNETOS

the same magnet frame, one armature being displaced angularly with relation to the other. Ordinarily for a two-cylinder motor the armature would be driven at one-half crankshaft speed and the angular displacement of one armature relative to the other would be equal to half the angle between cylinders. This solution of the problem is not a very fortunate one, however, as a double armature magneto is expensive, and on these small two-cylinder motors it does not pay to use such expensive ignition apparatus.

Bosch Magneto for V Motors—A very neat solution of the problem of magneto ignition for two-cylinder V motors has been worked out by the Bosch firm. The magneto runs at one-

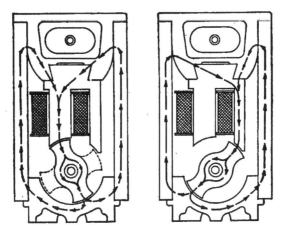

FIG. 126—MAGNETIC CIRCUIT DIAGRAMS OF TEAGLE INDUCTOR MAGNETO.

half crankshaft speed and therefore must produce sparks at intervals corresponding to 180 + a/2 degrees and 180 — a/2 degrees of armature motion. The peaks of the armature currents must be spaced the same.

As shown in Fig. 127, the two pole shoes of the shuttle armature are displaced longitudinally with respect to each other, one shoe extending beyond the central part of the shuttle at one end, and the other shoe at the other end, the inner end surfaces of the two shoes lying in the same median plane of the magneto. The pole pieces of the magneto also change in section at the middle of their length. Referring to Fig. 128, which represents a front elevation of the magneto partly

SPECIAL TYPES OF MAGNETOS

in section, the front half of the left-hand pole piece subtends a comparatively small arc of the circumference, while the rear half subtends a considerably greater arc. Correspondingly, the front half of the right-hand pole piece subtends a comparatively large arc of the circumference and the rear half a small one.

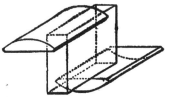

FIG. 127—ARMATURE CORE OF BOSCH MAGNETO FOR TWO CYLINDER V TYPE ENGINES.

In the drawing the halves of the pole pieces and the armature core toward the reader are cross sectioned. The voltage induced in the winding of the armature attains its maximum when the pole shoes of the armature are midway between the pole pieces of the field, as then the direction of the magnetic flux through the armature core is reversed and its rate of change is a maximum. It will be seen that the armature is shown in one of these positions of maximum voltage. The next position of maximum voltage is reached after an angular motion of $180 + a/2$ degrees, as indicated in dotted lines. Then both armature pole pieces are again midway between the field poles at their respective ends of the magneto. From the moment of this maximum voltage to that of the next an angle of $180 - a/2$ degrees must be passed through by the armature. In Fig. 129 the curve in full lines represents in a general way the voltage wave obtained from a magneto of this type, while the dotted curve represents the voltage wave of a standard magneto.

High Tension Insulators—All parts of the magneto carrying the high tension current must be covered with material of high insulating properties (dielectric strength). Some of these parts at least are of complicated form and can be fashioned only by molding. The materials used, moreover, should preferably have considerable mechanical strength and not be injured by moderately high temperature. Hard rubber has the

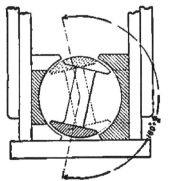

FIG. 128—MAGNETIC CIRCUIT OF BOSCH MAGNETO FOR V TYPE ENGINES.

162 SPECIAL TYPES OF MAGNETOS

necessary insulating qualities but does not withstand high temperatures. Such parts as distributer blocks are generally molded from insulating composition possessing the various qualities required. Of such compositions there are several, such as bakelite and stabilit. The former is an American, the latter a German product. Bakelite is a synthetic product which is solid at ordinary temperatures, fuses at moderately high temperatures and becomes infusible and insoluble when subjected to high temperatures. The parts of this material are formed in molds held on steam jacketed heads in a press, the steam being at about 100 pounds per square inch pressure, which corresponds to a temperature of 340 degrees Fahr. At this temperature the composition flows readily in the molds. A molecular transformation takes place when the composition

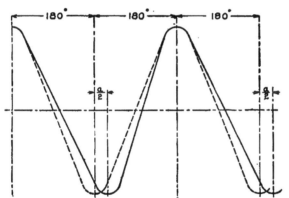

FIG. 129—VOLTAGE WAVES OF V TYPE AND STANDARD MAGNETOS.

is subjected to high pressure in the heated mold. It is allowed to cool under pressure in the mold, is then removed, and finally is subjected to a special treatment consisting in placing it in a chamber containing an air pressure of from 100 to 120 pounds per square inch, which is heated by live steam under 40 to 100 pounds pressure in a jacket surrounding the chamber. This final treatment gives the product a permanent character. Bakelite thus treated withstands temperatures up to 600 degrees Fahr., is insoluble and not affected by atmospheric influences. Most of these insulating compounds are the products of a reaction between phenol and formaldehyde. There is, however, a German product used for this purpose which is made from skimmed milk and known as galalith (milk-stone).

SPECIAL TYPES OF MAGNETOS

Galalith is manufactured from casein by means of formaldehyde. A solution of casein is obtained by treating skimmed milk with caustic alkali. The solution is clarified and the casein precipitated by means of acids and then filtered. The water is then partly extracted by pressure and the product dried very slowly. The drying process extends over a period of several weeks. The casein plates thus obtained are thoroughly saturated with formaldehyde and dried again. The product is somewhat transparent, of a yellowish-white color and very similar to horn.

Galalith is an excellent insulating material and may be fashioned either in the cold state or after it has been softened by using hot water. It is free from odor and is not so inflammable as celluloid, but is never entirely transparent and it is not possible to manufacture it in very thin sheets. Its specific weight is 1.317 to 1.35; the hardness is 2.5 according to the Moh scale.

In the case of high-tension distributers the metal parts are sometimes molded in rings or blocks of hard rubber and these are then molded in some synthetic insulating material.

Care and Adjustment of Magneto—The modern ignition magneto requires very little attention and is undoubtedly one of the most dependable machines ever designed. Of course, it must be oiled now and then. All magnetos have ball bearings on the armature shaft and some have this type of bearing also on the distributer shaft, although the latter carries exceedingly little load. Ball bearings require lubrication only at long intervals and the object of the oil in their case is not so much to form a separating film between journal and bearing as to protect the balls and races against rust. There is a natural inclination to use cylinder oil, as this is always at hand in the garage, but this oil has a tendency to gum. It is better to use a lighter grade of oil, because there is neither great pressure on the bearing surfaces nor are they subjected to high temperatures. Oil of the grade generally used for sewing machines is best adapted for this purpose. A few drops every 1000 miles of running is generally sufficient. Care must be taken not to over-oil the magneto, as, if oil should get into the distributer, it is apt to cause trouble. Oil on the interrupter points also is likely to lead to trouble, causing sparking at the points and irregular operation.

Magnetos should always be kept clean and as free from moisture as possible, because dirty water is a conductor of electricity, and there is always some chance for it to work through joints and crevices to the high tension conductors and as a result to cause leakage of current. The distributer plate must be removed occasionally to see whether any carbon dust wearing off the distributer brush has collected on it. This dust may form a leakage path between distributer segments and thus cause irregular firing. If any dust is found it should be removed by wiping the plate with a cloth, and if the carbon is caked the cloth may be moistened with gasoline to render it more effective. After the plate has thus been cleaned a light film of oil may be applied to its wearing surface to prevent excessive wear of the segments and carbon brush. In low tension magnetos the carbon brush which takes the current from the collector ring may fail through sticking in its holder or the ring becoming gummy, thus preventing good electrical contact between brush and ring. The remedy for this trouble is to refit the brush in its holder and to clean the ring with a fine grade of sand paper.

Adjustment of Interrupter—Undoubtedly the most important part of the magneto so far as maintenance is concerned is the interrupter. If the magneto fails to furnish a spark this part should be carefully examined. The contact screw may require adjustment to obtain the best length of opening between contacts, usually about 1/64 inch. If the platinum contact points are rough or in bad condition they should be carefully dressed with a fine flat file. It is well to observe whether the interrupter lever moves freely on its pivot, and in case it does not do so it may be necessary to ream out the hole through it. All screws and nuts in the interrupter should be examined as to their tightness and screwed up tight if necessary. As to whether the fault is in the plugs and cables or in the magneto itself, in case of irregular firing, can be observed by removing one of the high tension cables at a time and observing the spark at the safety spark gap. It is not advisable, however, to let the spark play at the safety spark gap for any great length of time, for this spark requires a higher voltage than a spark at the plug, and therefore subjects the insulation of the secondary winding to higher strains.

Location of Troubles—One thing that it is well to keep in mind in case of ignition trouble is that missing of one cylinder is almost certain to be due to defects in the spark plug or the cable leading to it, while an absolute failure of the ignition is due to a defect in the low tension circuit, which includes the interrupter. The wire leading to the switch may have become grounded or the switch itself short-circuited. If the engine fails to stop when the spark is turned off—provided this is not due to self ignition caused by overheating of the cylinders—the trouble may be due to failure of the switch members to make proper contact, to a break in the switch wire or to a loose connection in the switch circuit. All carbon brushes, of course, wear out eventually and have to be replaced.

If an armature should be damaged so that it will not generate, it should be sent to the makers for repairs, or at least to some firm making a specialty of magneto repairs. In case a magneto is entirely dissembled, a keeper, that is, a heavy plate or bar of soft steel or iron, must be applied to the poles of the magnets, or they will lose some of their strength. In regular operation the magnets of a magneto having such a good magnetic circuit as the average two-pole magneto seldom lose their strength, unless they are subject to very high temperature by proximity to the exhaust manifold or there are other unusual conditions. Care must also be used in reassembling the magnets so as not to reverse one of them. In many cases where complaints are made to manufacturers that the magnets have lost their strength it is found that they have been taken apart and in reassembling one magnet has been turned the wrong way so that it opposes the others. This, of course, is most likely to occur with magnetos comprising a considerable number of magnets.

CHAPTER XVI

Methods of Spark Timing

One of the undesirable features connected with conventional magnetos in which only the moment of opening the primary circuit is varied in timing the spark, is the small range in timing available, usually only about 25 degrees. This is due to the fact that there is an appreciable current flowing in the armature only during the time of a small angular movement of the armature. Fig. 130 shows the approximate form of the voltage impulse induced in the armature. Owing to the self-induction of the armature the current wave will lag slightly behind the electromotive force wave, attaining its maximum value 4 or 5 degrees later than the former. Of course, the most voluminous spark is produced in the secondary if the primary circuit is broken at the moment the current in it is at a maximum, that is, when the central plane through the pole pieces of the armature has passed the vertical by 4 or 5 degrees. If the opening of the primary circuit occurs either earlier or later the volume of the spark will thereby be reduced, and if it occurs sufficiently earlier or later the secondary electromotive force will not be sufficient to produce a spark. The possible advance and retard are therefore limited to a short range ahead and behind the current peak position. In magnetos in which spark timing is effected by varying the time of opening the primary circuit relative to the time of maximum induction in the armature, the timing range is seldom in excess of 25 degrees of crank motion, unless the pole pieces are given a special shape.

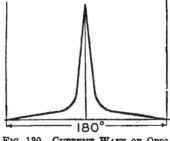

FIG. 130—CURRENT WAVE OF ORDINARY MAGNETO.

METHODS OF SPARK TIMING

Reason for Peaked Current Curve—It may be of interest to analyze the reason for the peaked form of current wave, more particularly since a quite different form, the sine wave, was described in an earlier chapter as characteristic of alternating currents. The peak form of wave is due chiefly to the H type armature. When the armature core is in its position of rest, that is, with the armature poles facing the magnet poles, a slight motion of the armature makes almost no change in the number of lines of force passing through the armature winding. In fact, the armature may move through a certain angle without effecting any increase in the reluctance of the magnetic circuit, as the armature poles remain masked by the field poles. As the armature poles move out from under the field poles the area of the air gap becomes less, but at first the proportional reduction of this area for an armature

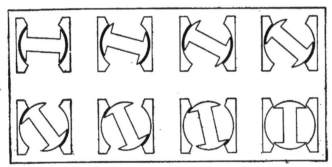

FIG. 131—ILLUSTRATING CHANGE IN MAGNETIC RESISTANCE OF CIRCUIT WITH SHUTTLE TYPE ARMATURE.

motion of, say, 10 degrees, is small. But when the trailing edge of the armature pole approaches the far edge of the field pole, the area of the air gap decreases at an exceedingly rapid rate; consequently the magnetic flux decreases very rapidly and a high e.m.f. is induced in the armature winding. In Fig. 131 the armature core and field poles are shown in eight different positions, succeeding positions being separated by equal angles of armature motion. It will be seen that the proportional decrease of the air gap, which is measured by the length of the heavy black sector, is at first very small and toward the end very large.

In order to widen the peak of the current wave—in other words, to increase the angular range through which the armature current is sufficiently strong to produce a spark—the magneto pole pieces must be so changed that the reduction in the number of lines through the armature as the armature core approaches the vertical position occurs more gradually. One way of accomplishing this is to give the magnet poles a helical or twisted form, so that the trailing edge of the armature pole, instead of passing out from under the magnet pole all at once does so gradually. In Fig. 132 is shown an end view of a magnet frame with helical pole pieces and a development of the pole faces on a plane surface. The change in the magnetic flux here again is most rapid when the armature pole is located symmetrically between field poles, but since the trailing edge of the armature pole begins to pass out from under the field pole considerably earlier than in the standard design of magneto, the decrease in the magnetic flux through the armature core is more "spread out" and hence the current curve is flattened. Another effect of giving the pole shoes a helical form is said to be to reduce the demagnetizing

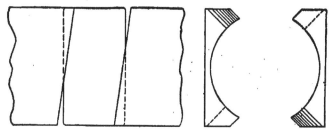

FIG. 132—HELICAL EDGED POLE FACES.

effects on the magnets. This principle was first employed by the Eisemann Magneto Co., which now gives its pole shoes a somewhat different form, evidently intended to accomplish the same object. This may be described as follows: The edge of the pole face, starting from one end, instead of running parallel to the axis of the armature bore, runs helically around it for half the length of the bore, and thence it runs helically in the opposite direction to the other end. In other words, the adjacent edges of unlike poles come closest together at the middle of their length and recede uniformly from each other from the middle towards both ends.

A different plan for obtaining a comparatively flat current wave is to make the pole edges comb-shaped, or, in other words, to saw into the pole edges, as shown in Fig. 133. With the ordinary pole construction the magnetic flux through the armature remains very high until a moment before armature and magnet pole part company, but by reducing the effective

METHODS OF SPARK TIMING

axial length of the pole area by about one-half, by means of the saw slots, the angle of a high rate of reduction in the flux, and consequently of heavy armature current, is increased. Combed pole pieces of this general type are used by the Bosch Magneto Co. In addition to being "combed" the pole shoes are also made unsymmetrical.

Rocking Magnet Frame for Spark Timing—It has been pointed out repeatedly that, other things being equal, the hottest spark is produced if the primary circuit is interrupted when the current in it is at a maximum. But with the ordinary method of spark timing the moment of interrupter action is varied relative to the moment of maximum induction, hence the spark will be stronger for certain positions of the spark lever than for others. Naturally, the timing gear would be so

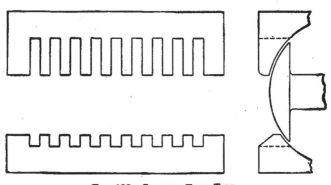

FIG. 133—SLOTTED POLE TIPS.

adjusted as to cause a hot spark to be produced with the timing lever in the position corresponding to normal engine speeds. This means that when the engine runs at low speed, for which the spark lever must be set to give late ignition, a weak spark is produced. Since under these conditions the engine generally receives a rather poor charge, which is difficult to ignite, it would be much better if the spark could be maintained at its maximum value. Various expedients have been resorted to with a view to this end. One of the first of these was that employed by the makers of the Mea magneto who, in order to change the timing of the spark, rock the magnet frame around the armature axis. To make this easier, a bell-shaped magnet is used which is mounted on trunnions in a housing, with the axis of the bell coinciding with the armature axis. It will be readily understood that if the magnet frame is moved around in the direction in which the armature rotates, the armature will reach the position of maximum induction somewhat later. Also, since the so-called stationary contact of the interrupter is fixed to the magnet frame, it will be moved around through the same angular distance as the frame, and the interruption of the primary circuit will be retarded just as much as the instant of maximum current. Hence, if the interrupter is once set to break the circuit when the primary current is at its maximum, it will always break it at that point, whether the spark is advanced or retarded, and the volume of the spark, consequently is not affected by advancing or retarding the spark.

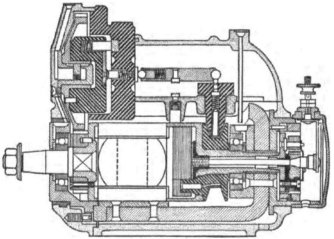

Fig. 134—Mea Magneto Mounted on Trunnions for Spark Timing.

On the contrary, the spark always has its maximum value. Spark timing is effected by rocking the whole magnet frame in its supporting trunnions. Turning it in the direction of armature rotation retards the spark and turning it in the opposite direction advances the spark. The same plan is used by Eisemann on some models embodying field magnets of conventional form, and by Splitdorf in the Dixie magneto in which, instead of the field magnets, the armature core with its poles is rocked around the rotor axis. The Mea magneto is shown in section in Fig. 134. The magneto proper is enclosed and supported in bearings in a separate housing and for the sake of contrast the section of the housing is shaded.

METHODS OF SPARK TIMING

Spark Timing by Helical Slots—Instead of rocking the field magnet around the armature axis, we may move the armature around that axis and relative to its driving shaft to attain the same final result. Suppose the armature to be in driving connection with the engine crankshaft and the interrupter to be in the position where the break is about to occur. If now we disconnect the armature shaft from its driving shaft, swing the armature back through a certain angle and connect its shaft again to the driving shaft, then the circuit naturally will be broken later in the cycle of the engine and the spark will be retarded. In practice this motion of the armature shaft relative to the driving shaft is effected by means of a sliding sleeve with a pin on it. One of the two shafts to be connected is hollow and passes over the other, and the sliding

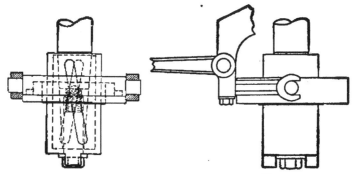

FIG. 135—HELICAL GROOVE AND KEY TIMING MECHANISM.

sleeve, which is outside of both, carries a pin which extends through a helical groove in the hollow shaft and into a straight or helical slot in the inner shaft. If the slot is helical it must be curved in the opposite direction from the groove in the hollow shaft. The sliding sleeve is provided with a circumferential groove on the outside into which engages a forked lever, and by sliding the sleeve lengthwise in qne direction or the other the magneto armature can be moved angularly with relation to the driving shaft. Fig. 135 shows such a construction in which two pins and two sets of slots are used.

Automatic Timing—Instead of shifting the sliding collar by hand it may be shifted automatically by means of a centrifugal governor, and this principle is employed by Eisemann. Three sectional views of the Eisemann automatic timing

172   METHODS OF SPARK TIMING

mechanism are shown in Fig. 136. Driving shaft A is cut with a helical key and is surrounded by a sliding block having a corresponding helical keyway. Sliding block B is adapted to slide in the guide CC which is secured to and forms an ex-

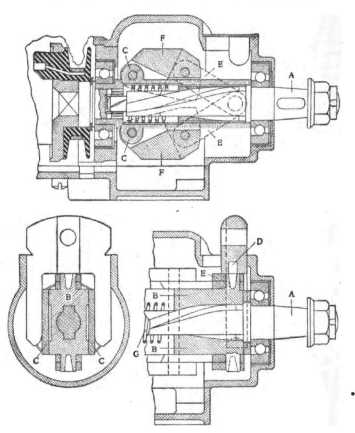

FIG. 136—EISEMANN TIMING GOVERNOR.

tension of the armature shaft. Sliding block B is provided near its outer end with a pair of trunnions D to which connect two pairs of links E that are hinged to the free ends of the centrifugal weights F. These weights are pivoted to the guide C. A coiled spring G normally keeps the sliding block B against a collar on the driving shaft and the centrifugal weights close to the axis of rotation. This is the position of maximum retard. As the engine speeds up the centrifugal force acting on the centrifugal weights causes them to move out from the axis of rotation, drawing sliding block B toward the armature against the pressure of spring G, and through the intermediary of the helical key and keyway, angularly moving the armature relative to the driving shaft in the direction of armature rotation, thereby advancing the spark. The spark is advanced in direct proportion to the speed, and this is effected entirely automatically without attention from the driver.

Automatic Timing Coupling—A magneto coupling incorporating a spark governor is manufactured by Herz & Co. (Fig. 137). The coupling consists essentially of two discs, which have spiral grooves cut in their adjacent faces, the grooves

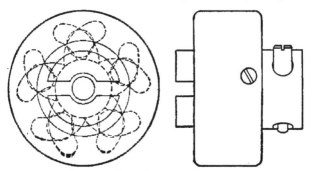

FIG. 137—HERZ GOVERNOR COUPLING.

in one disc being right-handed and those in the other disc left-handed. There are five grooves in each disc, and in the pockets formed where two grooves cross are located steel balls. The driving effort transmitted through the coupling tends to keep the balls in a position close to the shaft, but as the speed of the coupling increases, centrifugal force acting on the balls causes them to move outwardly in the grooves, thereby angularly advancing the disc secured to the armature shaft relative to the disc secured to the driving shaft. This governor coupling is very compact and can be used in place of the ordinary magneto coupling. It permits of any timing range desired, by varying the inclination of the grooves, but the standard construction provides for a range of 40 degrees. All parts of the governor are made of tool steel, hardened and ground.

Automatically Timed Battery Ignition—It has already been pointed out that with a so-called closed circuit type interrupter the moment of interruption of the primary circuit is not affected by the engine speed. Such interrupters or timers are also known as synchronous timers. Open circuit interrupters, as a rule, are non-synchronous and require a greater degree of advance for a certain increase in engine speed, for the reason that this advance must compensate not only for the lag due to flame propagation, but also for the lag due to interrupter action. For this reason, the spark, if hand timed, re-

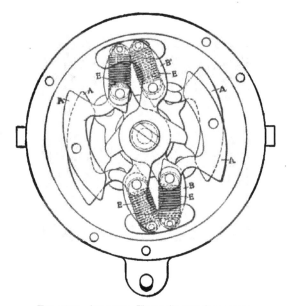

FIG. 138—ATWATER KENT SPARK GOVERNOR.

quires more constant attention with an open circuit than with a closed circuit system, and automatic timing is most extensively used with the former.

In connection with the Atwater Kent interrupter, already described, there is used an automatic timer or governor which is located at the bottom of a housing that also accommodates the interrupter and the high tension distributer. A plan view of the governor is shown in Fig. 138. The device is of symmetrical design and absolutely balanced. It comprises four governor weights A, and two double armed brackets BB. The

METHODS OF SPARK TIMING

former is connected to a short driving shaft entering the igniter housing centrally through the bottom, while the latter is connected to the notched interrupter shaft which is slotted across its end for connection with the distributer shaft. The two governor weights on the same side of the shaft are pivotally connected together at their middle, and each has a pivot support on one of the bracket arms. Four coiled springs E extending between the pivot pins of the governor weights and the opposite bracket arms tend to keep the weights in a posi-

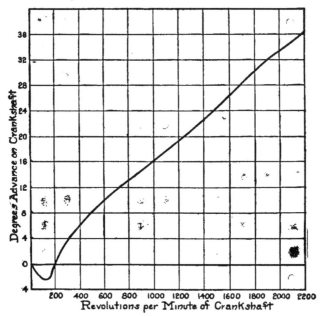

FIG. 139—SPEED-SPARK ADVANCE CURVE OF ATWATER KENT SYSTEM.

tion close to the shaft, but when the shaft rotates, centrifugal force tends to move the weights outward, and any outward motion of these weights is accompanied by a rotation of the notched shaft with respect to the driving shaft. The motion of the governor weights is radially out from the centre, and the centrifugal force on them therefore acts to the best advantage.

The governor thus moves the interrupter shaft ahead of the driving shaft as the speed increases, thereby advancing the ignition. Fig. 139 is a diagram showing the advance corresponding to different speeds of rotation.

Insurance Against "Back-Kicks"—An advantage of automatically-timed ignition is that the danger of back-kicks in starting the motor is obviated. When hand cranking was practiced exclusively, many drivers suffered broken wrists as a result of forgetting to set the spark late before cranking the motor, thus causing the charge to be fired before the piston in the firing cylinder had reached the top of the stroke. Strictly speaking, dangerous back-kicks are possible only with battery ignition, because with magneto ignition, in order to generate a spark, the crank has to be turned vigorously and the momentum of the fly-wheel will then carry the engine over dead center even if the charge is fired before the dead center position is reached. With a battery system the crank is usually just pulled over dead center, as nothing is gained by cranking fast, and if the spark lever happens to be set for early ignition the operator runs grave risk of personal injury. Such risks are obviated by the automatic advance or spark governor.

Spark governors are also used by the Westinghouse Electric & Mfg. Co. and the Dayton Electrical Laboratories Co. in their battery ignition systems.

Various expedients have been resorted to in order to prevent the cranking of an engine while the spark lever is in the advanced position. One consists in a sort of block which is moved between the engaging ends of the crankshaft and the starting crank, by means of the spark lever, in the act of setting the spark early. It is then impossible to engage the starting crank. Another consists in the provision of a mechanism whereby the act of pushing in the starting crank for engaging it with the claw on the crankshaft automatically retards the spark.

CHAPTER XVII

Combined Magneto and Battery Systems

In Europe it is the general practice to depend for ignition solely on the high tension magneto, in starting the motor as well as in regular operation. Starting a motor by hand on magneto ignition requires rapid spinning of the crank, instead of merely turning the crank over the dead center. Spinning is readily accomplished with a motor of small cylinder dimensions, but is quite difficult with motors of such large cylinders as were in common use in this country previous to 1910. This condition led to the widespread adoption in this country of a battery as an auxiliary source of current, which is used in conjunction with more or less of the apparatus of the magneto system. Such ignition systems employing a double source of current are usually referred to as dual systems.

The battery is most easily combined with a magneto and coil system, and an example of this dual system (the Remy) was illustrated in Fig. 109. To render a magneto and coil system dual requires no additional apparatus besides the battery, which is connected to what would otherwise be the dead button of the ignition switch and to the secondary grounding terminal of the coil. With a high tension magneto the problem is somewhat more complicated and affords opportunity for a variety of solutions which are perhaps best illustrated by the various combined magneto and battery systems offered by the Bosch Magneto Co.—the "Two Independent," the Dual and the Duplex systems.

"Two Independent" System—With the "Two Independent" system two sets of spark plugs are used and the engine can be fired either by sparks from the battery, by sparks from the magneto or by both acting simultaneously. The battery system comprises a ccmbined interrupter and distributer, a coil and a separate set of spark plugs. The interrupter is of the magneto type and therefore gives a single spark per explosion. Combined with the coil is the ignition switch which serves for both magneto and battery ignition. The coil is

178 COMBINED MAGNETO AND BATTERY SYSTEMS

designed to be carried on the dashboard in a horizontal position, extending forward from the dash, while the switch is on the rear side thereof. Referring to Fig. 140, which is a sectional view of the coil and switch, the movable brass cover A of the coil carries the switch handle B and is secured to the cylindrical housing C by means of a bayonet joint. A pin set on the coil end plate engages an opening in the cover which causes the coil and the cover to move together. Switch contacts are located on the other end plate of the coil, and permit the operation of the switch by the movement of the

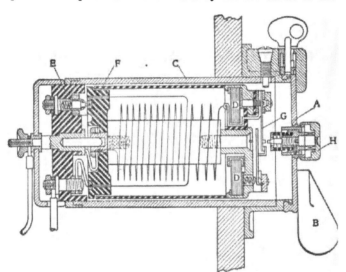

FIG. 140—SECTIONAL VIEW OF BOSCH HORIZONTAL COIL, SWITCH AND STARTER BUTTON.

cover. There are four switch positions, viz., "Off," "Battery," "Magneto and Battery" and "Magneto," indicated on the switch cover by their initial letters. The base of the coil housing is formed by the stationary contact plate E and the contacts carried on it register with contacts on the movable switch plate F. Partial rotation of the coil by the movement of the cover plate causes the different switch contacts to engage. One end of the primary winding is connected to a segment on switch plate F and the other leads to a vibrator G at the rear end of the coil. The condenser D for this vibrator is located between the coil windings and the vibrator.

COMBINED MAGNETO AND BATTERY SYSTEMS 179

The coil is provided with a key-lock, which may be operated only when the coil is in the "Off" position. This prevents unauthorized use of the engine, and by making it impossible to lock the switch in any of the operative positions, renders it unlikely that the switch will be left unintentionally in one of the battery positions and thus allow the battery to run down. With a vibrator coil system it is extremely unlikely that the engine be left with the switch closed and the primary circuit completed through the timer so that the battery would discharge through the coil, as the vibrators in that case give audible warning. With a magneto type interrupter, however, there is no such warning, and if such an interrupter is used on a battery there is always some danger that the switch may be inadvertently left closed and the battery will run down. Evidently, where a lock is provided on the switch and this lock is habitually used, this danger is practically eliminated.

Starting on the Spark—One advantage of battery and vibrator coil ignition over ignition by magneto only, and some other forms of battery ignition, is that with the former a spark may be produced in an engine while the latter is at rest, and this often enables one to start an engine with four or more cylinders without cranking. This phenomenon may be explained as follows:

A four cylinder engine usually comes to rest with the pistons in all cylinders near midstroke. An engine usually stops after ignition is cut off, because there is not sufficient energy left in the flywheel to carry it over one of the following compressions. The piston in the cylinder in which compression takes place may get near the top of the stroke but stops short of the end of it. The compression pressure acting on the piston in this cylinder will then force the piston back some distance. Of the other three cylinders, the two which were on their inlet and exhaust strokes, respectively, have their valves open and are therefore without pressure. On the other hand, the cylinder which would have had a burning charge in it had the ignition not been cut off, has as much combustible charge in it as the one on its compression stroke, and, besides, its valves are closed. Therefore, when the crank reverses, the piston in this cylinder will move up and recompress the charge, and when it is at midstroke the pressure on this piston and that on the one which failed to fully compress its charge are equal, and the crankshaft is in equilibrium. Therefore, the engine comes to rest in this position.

Now, the contact segments of a timer usually extend over more than one-half the circumference, and if the spark lever is set for late ignition the movable contact of the timer will still be on the segment corresponding to the spark plug in the cylinder in which in normal operation the firing stroke would be performed, when the piston in this cylinder reaches midstroke. Therefore, when the switch is closed the primary circuit will be closed, the magnetic vibrator will begin to operate and a shower of sparks will be produced in the cylinder referred to. If the engine has not been at rest for any great length of time and its cylinders are fairly tight, there will be an explosive charge in the cylinder in which the sparks occur, which will be exploded, and the resulting pressure on the piston will carry the crank around, thereby drawing in a fresh charge in one of the other cylinders, compressing and firing it. This is known as "starting on the spark," and this method of starting is available on all engines having an ignition system comprising a magnetic vibrator. Of course, one cannot depend upon this method for starting under all conditions, but since it often relieves the driver of the work of manual cranking it is a valuable feature. In order to render starting on the spark more certain it is well, in stopping the engine, as the spark is shut off, to open the throttle wide. This insures a rather rich mixture being drawn in during the last few strokes of the engine, and what is necessary to make the starting spark effective is a good, strong mixture in the cylinders.

The Bosch Two Independent system employs a magneto type interrupter for the battery circuit, but to permit of starting on the spark, a magnetic vibrator G is provided in addition. Referring to Fig. 140, the stationary contact of this vibrator is carried on a button H, located centrally in the switch plate. In an ordinary magnetic vibrator the two contact points are normally together, so that as soon as the ignition switch is closed the primary circuit is completed and the vibrator begins to operate. In the vibrator here shown, the two contacts are normally apart so that the vibrator cannot operate. But by pressing button H in against a stop, its contact can be brought up against the contact on the vibrator blade G, and as this closes the primary circuit the vibrator begins to operate and causes a shower of sparks in the engine cylinder. Of course, the button is effective only when the switch is in either the "Battery" or "Magneto and Battery" position, that is, when the battery is connected to the coil. The vibrator is connected in parallel with the mechanical interrupter and therefore operates only when the circut is open at the latter, for when the circuit is closed in the interrupter the magnetic vibrator is short circuited. When the pistons are at midstroke (the usual stopping position) the

COMBINED MAGNETO AND BATTERY SYSTEMS 181

circuit through the interrupter is always broken. The starter button may be fixed in the working position by pressing down and rotating it slightly, so that even if the motor will not start on the spark the advantage of a series of sparks may be obtained for cranking the motor.

Switch Connections—Referring to Fig. 141, it will be

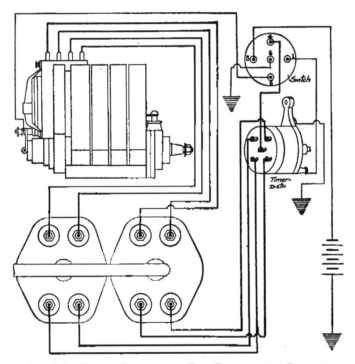

FIG. 141—WIRING DIAGRAM OF TWO INDEPENDENT SYSTEMS.

noticed that there are one central and four radial terminals on the back of the switch, the central one being the ground terminal. When the switch is in the "Off" position the battery circuit is broken and the magneto is grounded (switch contacts 6 and 2 connected), hence no sparks can be produced if the motor is turned over. If the switch is turned to the next or "Battery" position, the magneto remains grounded but

182 COMBINED MAGNETO AND BATTERY SYSTEMS

the battery circuit is now completed (switch contacts 5 and 1 connected), consequently sparks may be produced in the cylinder by pressing the starting button or by turning over the engine so as to operate the mechanical interrupter. If the switch is turned to the "Magneto and Battery" position, the only change made in the connections is that the ground of the magneto is now broken, and the magneto, too, produces sparks. Turning the switch to the final or "Magneto" position opens the battery circuit and results in ignition being furnished by the magneto alone.

Dual System—The Dual system differs from the "Two Independent" system in that only one set of spark plugs and only the magneto distributer is used, and the interrupter for

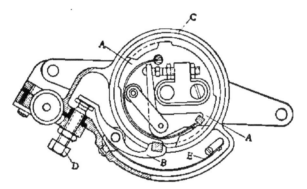

FIG. 142—DOUBLE INTERRUPTER FOR DUAL SYSTEM.

the battery current is combined with the interrupter for the magneto current. This double interrupter is shown in Fig. 142. The central portion of the magneto interrupter is exactly the same as on an ordinary magneto as illustrated in Fig. 115. In this case, however, the interrupter base (Y in Fig. 113) is provided with two cam lobes AA on its circumference which actuate the breaker arm B pivotally supported on the interrupter shell C. Arm B carries a contact point adapted to make contact with the stationary contact screw D which is supported in the housing but insulated therefrom. The two contacts are held together by the spring E, except when the fibre shoe of arm B rides on cam lobes AA. Since arm B is grounded when the contact points are together the

COMBINED MAGNETO AND BATTERY SYSTEMS 183

battery current coming in at terminal D flows through the contacts into ground.

Dual System Connections—A wiring diagram of the dual system is shown in Fig. 143. Coil and starting vibrator are exactly the same as in the case of the two independent systems, but the switch is arranged somewhat differently. It has only three positions—viz., "Magneto," "Battery" and "Off." As in this case the battery spark is also distributed by the

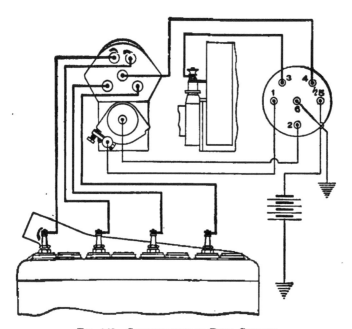

FIG. 143—CONNECTIONS OF DUAL SYSTEM.

magneto distributer, it is necessary to run two high-tension leads from the magneto to the switch, as against one in the previously described system. One of these, connecting to the collector brush at the driving end of the magneto, carries the magneto high-tension current to the switch, and the other connecting to the central terminal of the distributer, carries either the magneto or the coil high-tension current to the distributer, according to the way the switch is set. In the

"Off" and "Battery" positions the magneto armature primary winding is short circuited by the switch; in the "Magneto" position it is not. In the "Off" and "Magneto" positions the battery circuit is opened by the switch; in the "Battery" position it is closed. Besides, the switch connects the armature secondary to the distributer in the "Magneto" position and the coil secondary to the distributer in the "Battery" position. Considerable importance is attached to the fact that in this system the main wearing part of any ignition system—the interrupter—is duplicated, so that if either of the two interrupters fails to work that will not incapacitate the engine.

Duplex System—The Bosch duplex system differs from the dual system in that only the source of current is duplicated, and all other parts serve both for magneto and for battery ignition. This system evidently is simpler and less expensive than either of the two already described. No separate "step up" coil is used, the battery current being transformed or "stepped up" in the magneto armature.

There are three conditions of operation with this system, as follows: (1) The engine may be started on the spark by means of a "press button"; (2) the engine may be started or normally operated with the switch in the "Battery" position; (3) the engine may be run with the switch in the "Magneto" position.

One difficulty that is encountered when it is attempted to use the magneto armature for transforming the battery current is that as soon as the engine begins to turn, while the battery current is still flowing through the armature winding, an induced current also appears in this winding. The latter is an alternating current, and since the battery current is continuous, every other impulse of induced current is opposed in direction to the battery current and tends to neutralize it. This obviously would not do, and, therefore, the battery current, before it flows into the armature, is "commutated" into an alternating current. As shown in Fig. 144, on the inside of the interrupter cover plate there are two nearly semicircular segments AA, which connect with binding posts on the outside of the cover. Against these segments bear two brushes B, C, carried on the interrupter base Y, of which one, B, is grounded and the other, C, connected to the insulated terminal I of the armature primary winding. Since the battery current comes in at one binding post and leaves at the other it will readily be seen that its direction of flow through the armature primary winding is reversed every half revolution.

The battery is connected in circuit with the switch and a small choke coil combined therewith. The reason for the

COMBINED MAGNETO AND BATTERY SYSTEMS 185

use of the latter is as follows: When the armature is stationary and no current is induced in its winding by motion relative to the field frame, the volume of the spark which can be produced by sending a battery current through its primary winding and then interrupting the circuit, is limited by the magnetic flux through the armature set up by this current. In the regular operation of the magneto the spark is not thus limited. Therefore, if it is desired to produce as powerful a spark for starting as is produced in regular operation (and this is essential), it is necessary to increase the store of magnetic energy set up by the primary current. This could be effected in one way by sending a greater current through the armature primary winding by using a higher

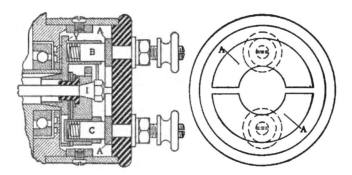

FIG. 144—COMMUTATOR FOR DUPLEX SYSTEM.

battery voltage, but inasmuch as the armature core is already nearly saturated it can be done more effectively by adding an extra core with a magnetizing coil somewhere in the circuit. A sectional view of the switch, choke coil and starter mechanism and a plan view of the latter are shown in Fig. 145.

A starting button is also used with this system, which in this case operates a mechanical vibrator instead of a magnetic one. Referring to Fig. 145, when the spring pressed plunger A is pressed in, its point moves the lever B, whose rounded end passes over the ridged fibre block C, secured to the contact spring D. Lever B is returned by coiled spring E when the plunger is released by the hand, and passes again over the ridged fibre block. It will readily be seen that when the plunger is fully pushed in there will be two

186 COMBINED MAGNETO AND BATTERY SYSTEMS

"makes and breaks" of the circuit during the instroke and two during the outstroke; in other words, four sparks are produced for every depression of the starting button. This vibrator, too, is connected in parallel with the magneto interrupter and operates only when the interrupter contacts are apart.

One respect in which the interrupter of Bosch magnetos for duplex ignition differs from those on independent magnetos is that the interrupter cams are longer, so that the interrupter is open a greater part of the time. Fig. 146 shows the interrupter open and the battery current now flows through the switch, choke coil, commutator, armature primary wind-

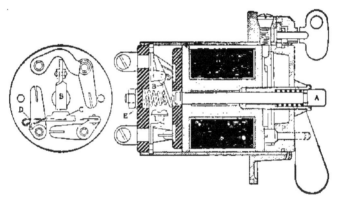

FIG. 145—CHOKE COIL, SWITCH AND HAND VIBRATOR OF DUPLEX SYSTEM.

ing, through ground to the interrupter lever, through the commutator and the hand vibrator back to the battery. When the current is interrupted by the hand vibrator the inductive effect on the secondary winding of the armature produces the sparks. When the magneto interrupter points close, the armature winding is short circuited, and the battery current then has two paths, through the armature primary winding and through the interrupter contact points. It will, of course, take the shortest path—viz., through the interrupter. As soon as the engine has picked up its cycle the switch is turned to the magneto position to save current.

Magneto Starting Helps—While they do not come strictly

under the heading of this chapter, we may here describe a couple of methods used to permit starting the engine on magneto sparks without "spinning" it. One system (Fig. 147) was patented by Robert Bosch and has been used in Europe for automobile and airplane work. It involves a separate magneto which in the case of an automobile is mounted on the dashboard convenient to the driver's seat. A crank handle connects through a set of speed multiplying gears to the arma-

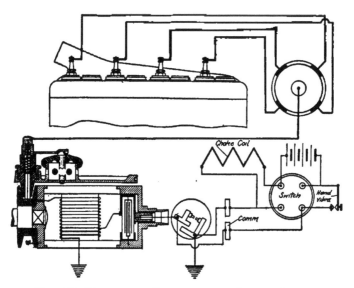

Fig. 146—Diagram of Connections of Duplex System.

ture. The magneto is a complete high tension machine except that it has no distributer.

Combined with the hand magneto is an automatic switch which has two positions, viz., starting and normal running. In the starting position this switch connects the secondary winding of the hand magneto armature with the cable leading to the distributer on the regular magneto, whereas in the normal running position the switch connects the high tension cable coming from the regular magneto collector ring brush to the cable leading to the distributer of the regular magneto.

188 COMBINED MAGNETO AND BATTERY SYSTEMS

While in the starting position this switch also grounds the cable coming from the collector ring brush of the regular magneto, so that as long as the hand magneto furnishes the spark the armature secondary winding of the magneto is short circuited. By turning the crank of the starting magneto, a powerful spark can be produced in the cylinder, and if there is a combustible charge in it this charge will be exploded and

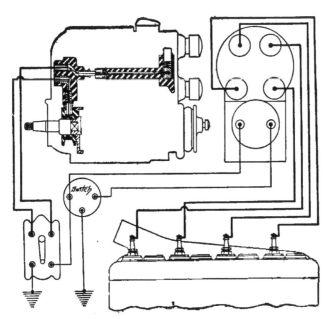

Fig. 147—Bosch Hand Magneto for Starting on the Spark.

the engine will start. This system, therefore, permits of starting on the spark without the use of a battery.

Impulse Starters—The other starting facilitating device was originated by Unterberg & Helmle and fitted to their U. & H. magneto. Two views of this device are shown in Fig. 148. A is the shaft to which the driving gear is keyed. Formed integral with this shaft is the steel disc B, which at one point of its circumference carries a pin C, over which passes the loop c of the coiled spring E. This is a spring

COMBINED MAGNETO AND BATTERY SYSTEMS 189

with large diameter spires of heavy wire. It will be understood that if the shaft A is turned in a right-handed direction, this spring E will be wound up. The other end of the spring is secured by a pin d to a brass disc F, which is fast to the armature. Disc F is of rather irregular shape, and is provided with a radial slot in which is located a steel ball G. When disc F moves in a right-handed direction the steel ball G abuts against a projection H of the stationary armature housing, and thereby prevents disc F and the armature from rotating. The result of this is that when shaft A is rotated the spring E is wound up; this continues until a conical de-

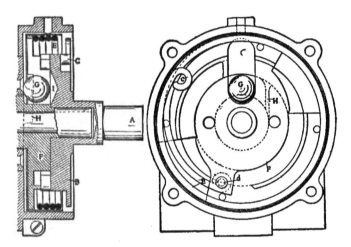

FIG. 148—U. & H. AUTOMATIC IMPULSE STARTER.

pression I in disc B comes opposite steel ball G. Then, the ball dropping into this depression, the armature is suddenly released, the spring distends and gives the armature a considerable speed of rotation, whereby a hot spark is produced. The projection or cam H is, of course, so arranged on the circumference of the field pole circle that the unwinding of the spring begins when the armature comes into the zone of greatest inductive activity. When the engine has run up to speed the steel ball is carried to the outer limit of the slot in disc F by centrifugal force and does not strike the projection H in its rotation.

190 COMBINED MAGNETO AND BATTERY SYSTEMS

Semi-Automatic Impulse Starters—A somewhat different type of impulse starter has come into wide use in this country on magnetos for farm tractor engines. These engines are often quite large and hard to crank and the operators in many cases are persons of limited physical strength. The prototype of this form of impulse starter, which must be set for starting and is therefore only semi-automatic, was the K-W illustrated in Fig. 149. Combined with the clutch member A is a spring case B in which there is a spiral spring C whose ends are anchored to the spring case and to the ratchet disk D respectively. The latter is fast upon the armature shaft to which it is pinned. A pawl E pivoted on the magneto frame

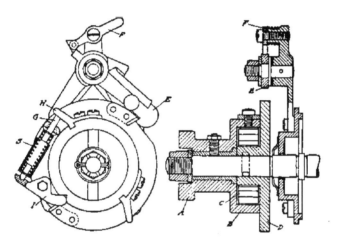

FIG. 149—K-W SEMI-AUTOMATIC IMPULSE STARTER.

is adapted to engage into the notches on the ratchet disk D. In the circumference of spring case B there are two knock-off cams H which, when the spring case is turned while the ratchet disk is being held stationary will force the pawl out of the notch on the ratchet disk and release the latter. When once released the pawl is held out of contact with the ratchet disk by lock F. As soon as the ratchet disk is released, it and the armature are spun rapidly by the spiral spring. Ratchet disk D jumps ahead of case B, with the result that the knock-off cam abutts against a plunger G mounted on a cushion spring S on the ratchet disk. This same spring S serves to press the dog I into a notch in the outer surface of spring case B. Before starting an engine fitted with this impulse starter, lock F must be pressed down so as to release pawl E. In the drawing pawl E is shown provided with hooks on opposite sides and spring case B with two knock-off cams H, the object being to permit of the use of the same impulse starter for magnetos turning either right-handedly or left-handedly. The ratchet holds the armature back for an armature motion of about 80 degrees.

CHAPTER XVIII

Two-Point Ignition

Reference has repeatedly been made to the fact that inflammation of the combustible charge is not instantaneous. The flame starts at the point where the spark takes place and is thence propagated throughout the mass of the charge. In consequence, the gaseous charge does not reach its maximum pressure instantly, as shown by the sloping explosion line of manograph diagrams taken at high speeds. Of course, the speed of flame propagation depends largely upon the composition of the combustible mixture. The greatest power would be obtained if the whole charge could be ignited instantly while the crank is in the dead center position. This is an ideal which is unattainable, but the nearer we can come to it the greater will be the power of the engine and the higher its fuel economy. Combustion lag becomes a rather important factor at very high engine speeds, and in a motor of the T-head type, in which, if the spark plug is screwed into one of the valve plugs, the distance from the spark terminals to the farthest part of the combustion chamber is unusually long. T-head motors were formerly largely used for racing, and in the endeavors to accelerate combustion in them the idea occurred that if two sparks were produced simultaneously in the combustion chamber at substantially opposite points, the power of the motor might thereby be increased. It seems that the plan was first applied in England in 1908. Dr. W. Watson in a paper read before the Institution of Automobile Engineers in 1910 mentioned that a certain four-cylinder

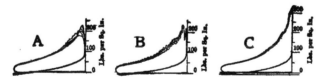

FIG. 150—MANOGRAPH DIAGRAMS OF ENGINE WITH SINGLE AND DOUBLE SPARK IGNITION.

TWO-POINT IGNITION

T-head motor of 3.4 inch bore and 4.8 inch stroke, with spark plugs over the valves, gave 18.4 i.h.p. on single spark and 20.8 i.h.p. on double spark at 1,100 r.p.m.; 26 i.h.p. on single spark and 29.2 i.h.p. on double spark at 1600 r.p.m. Fig. 150 shows three manograph diagrams taken from this engine by Dr. Watson, which clearly bring out the effect of two-point ignition All three were taken with the engine running at a speed of 1,600 r.p.m. In diagram A ignition was effected by means of a single spark in the inlet valve chamber; in diagram B by a single spark in the exhaust valve chamber and in diagram C by two sparks, one in each valve chamber.

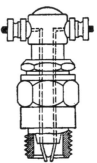

FIG. 151 — SPARK PLUG WITH BOTH TERMINALS INSULATED.

Two-Pole Spark Plugs—In order to be effective the two sparks must occur absolutely simultaneously, and this can be insured only by connecting the two spark plugs in series, so that the same spark passes the gaps of both. The easiest way to accomplish this with an ordinary ignition system is to use one ordinary spark plug and one plug in which both spark terminals are insulated. One such plug with both terminals insulated, the "Su-Dig," is illustrated in Fig. 151. It has two binding posts at the top and the spark plays between the ends of the two wires at the bottom, which are both insulated from the shell of the plug. The location of the two plugs in the cylinder and their connection to the coil is illustrated in Fig. 152. Coming from the coil, the spark first jumps the gap of the double pole spark plug, whence it flows to the other plug, jumps its gap and passes into the frame of the engine. Such plugs may be supplied with current from either a battery or a magneto.

Magnetos for Two-Point Ignition.—However, magneto manufacturers have developed a special type of magneto for two-point ignition which obviates the need for two different kinds of spark plugs. In this magneto, instead of one end of the armature secondary winding being grounded, both ends are insulated, and each is connected through a collector ring and brush to a high tension distributer, the magneto having two collector rings located side by side at the driving end and two high tension distributers arranged coaxially at the opposite end. Corresponding sectors of the two distributers are connected to the two spark plugs of one cylinder, the plugs being of the ordinary single pole type. Fig. 152 is a diagram of the secondary circuit of such a system. It will be seen that the spark passes through ground from one plug to the other and jumps from the insulated terminal to the shell in the first plug and the reverse way in the second one. In addition to accelerating combustion of the charge such an arrangement of the plugs evidently offers some insurance against misfiring due to sooting of the plugs, for should one plug be sooted and the other clean, the latter would act as an auxiliary spark gap.

It is obvious that for best results the spark plugs must be placed a considerable distance apart in the cylinder. The

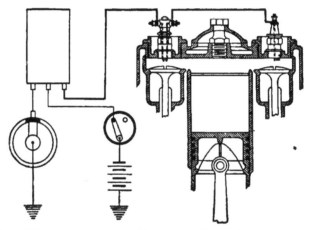

FIG. 152—ARRANGEMENT OF PLUGS AND THEIR CONNECTIONS FOR TWO-POINT IGNITION.

best position evidently would be such as would result in the least maximum distance from any point of the combustion chamber to either of the plugs. In practice, this ideal cannot be readily attained, but in a T-head motor placing the spark plugs in the plugs over the inlet and exhaust valves, respectively, comes very near it in effectiveness. T-head motors, moreover, are benefited most by two-point ignition, though the system is used also in valve in the head motors on racing cars.

It will be obvious from Fig. 152 that with two-point ignition the flame has to travel less than half as far as with single point ignition from one of the spark plugs in the figure. Consequently, combustion should be completed in much less time. This results in a higher explosion pressure, for the piston is still nearer the end of the stroke, and the combustion chamber, therefore, smaller, when the combustion is completed. The ultimate result of this is greater power and fuel economy. Looking at the matter from another standpoint, we may say that two-point ignition reduces the spark advance required.

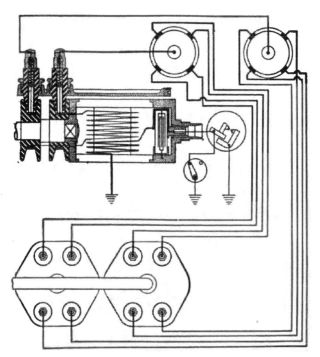

FIG. 153—DIAGRAM OF TWO-POINT MAGNETO IGNITION SYSTEM.

The spark advance, as has been explained, is made necessary by the combustion lag, and it is obvious that if the time required to complete the combustion is reduced the advance which must be given the spark is also lessened.

Experimental Results from Two-Point Ignition—A practical demonstration of the effect of two-point ignition on power and spark lead required was made in January, 1911, before the Society of Automotive Engineers by Otto Heins, president of the Bosch Magneto Co. The tests were made on a four-cylinder T-head motor of 3 9/16 inch bore and 4¾ inch stroke. A switch was used by means of which it was possible to fire the engine by either set of plugs (which were screwed into the valve plugs) or by both at the same time.

The greatest power output with single-point ignition with an advance of 45 degrees was 24-horsepower, and the same output was attained with two-point ignition with an advance of only 19 degrees. With two-point ignition the extreme power output with an advance of 32 degrees was 28-horsepower. This is an increase of 4-horsepower or 16.6 per cent over the best results with single-point ignition. Fig. 154

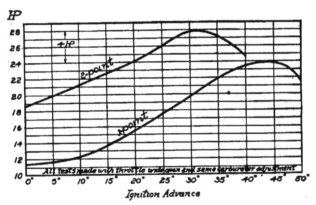

FIG. 154—INCREASE IN MAXIMUM ENGINE POWER WITH SPARK LEAD.

shows the increase in the maximum output of the motor with an increase in lead, for one-point and two-point ignition respectively, and Fig. 155 shows the horsepower-speed curves of the motor for one-point and two-point ignition respectively.

Practical Application of System—Two-point ignition, since it greatly reduces the advance required for maximum power, greatly lessens the objections to fixed ignition. For instance, it is quite possible to set the fixed spark for such an advance that the engine will give as much power as with a single spark with the most advantageous lead. Further, by using a switch permitting of using one or two-point ignition at will, the effect of a one-step spark advance can be obtained without the complication of the mechanism usually employed for obtaining the advance.

Distributerless Two and Four-Cylinder Magnetos

Distributerless Two and Four-Cylinder Magnetos—There is another system of magneto ignition in which two sparks are produced every time an explosion occurs, but they are not produced in the same cylinder. In fact, in this case the second spark serves no purpose whatever and is merely incidental. This form of ignition is used in two-cylinder engines in which both pistons start on the inward stroke together, and also in four-cylinder engines of standard construction. Four-cylinder engines are much more extensively used than two-cylinder, and the four-cylinder is therefore the more interesting case. In a four-cylinder engine of conventional design the two outer pistons move up and down together and the two inner pistons the same. Suppose one of the outer pistons to be at the end of its compression stroke, where the

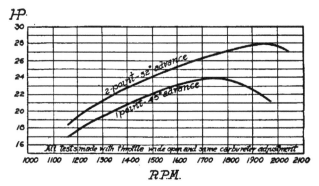

FIG. 155—POWER CURVES FOR ONE-POINT AND TWO-POINT IGNITION.

charge is normally ignited. Then the other outer piston is at the end of its exhaust stroke, so that there is only dead gas in its combustion chamber. Therefore, if we connected the spark plugs in the two outer cylinders in series, so that a spark occurred at both of them at the same time, the one in the cylinder whose piston was at the end of its compression stroke would fire the charge in that cylinder, while the spark in the cylinder whose piston was at the end of its exhaust stroke would have no effect. Moreover, since the pressure in the latter cylinder is substantially atmospheric, the fact that the spark has to jump the gap at the plug in this cylinder does not add materially to the resistance encountered by it.

The object in view in thus connecting spark plugs in two cylinders in series is to obviate the need for a high tension distributer on the magneto. Fig. 156 is a diagram of a high tension magneto ignition system for four-cylinder engines in which the high tension distributer is dispensed with. It will be seen that the armature secondary winding is not grounded, but has its terminals connected to two contact segments, forming part of a current collecting device carried at the driving end of the armature. The two segments are slightly displaced in a lengthwise direction and on opposite sides of the armature shaft. They are embedded in insulating material, and together with the latter form a pair of complete rings on each of which bear two oppositely arranged collector brushes. In the position shown, the high tension current evidently flows from one end of the armature secondary winding through one of the collector segments and its brush to the spark plug on one of the inner cylinders, and after jumping

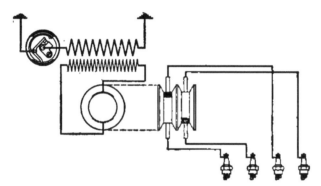

FIG. 156—FOUR CYLINDER MAGNETO IGNITION SYSTEM WITHOUT DISTRIBUTER

the gap of this plug it passes into the frame of the engine, then jumps the gap of the spark plug in the other inner cylinder, and returns through the second collector segment and its brush to the other end of the armature secondary winding. After the armature has made half a revolution, which corresponds to one piston stroke, its secondary winding is connected to the other pair of plugs, and sparks occur at these points.

With a system as here described it is well to limit the timing range after dead center, for if the spark lever is set to give late ignition, then the idle spark occurs not at the end of the exhaust stroke, but during the first part of the inlet stroke, and if the inlet valve has already opened and the spark plug is located over this valve, where it is directly in the path of the incoming charge, it is apt to ignite that charge and cause a so-called "back fire" through the carbureter. However, in most modern engines the inlet valve begins to open only about 15 degrees past dead center, and if the magneto is timed to give the proper maximum advance, with the timing range usually available it will not be possible to get a spark after the inlet valve has opened. At any rate, the system has been in use on two-cylinder motors for a considerable number of years, which shows that the feature mentioned is not a serious drawback in practical work. The object in view in the development of this magneto was to reduce the cost of construction, as compared with a magneto having a regular distributer.

CHAPTER XIX

Cables, Terminals, Switches and Wiring Methods

Ignition Cable—Special flexible cable is employed for wiring ignition systems. The cables consist of strands of tinned copper wire, the wires being arranged in concentric layers of which successive ones are stranded in opposite directions. The total effective cross sectional area of the wires for both primary and secondary cable is usually equivalent to that of a single No. 14 B. & S. gauge wire. As the individual wires are of very small size the cable will stand a great deal of twisting and bending without danger of breaking. The difference between primary and secondary cable lies mainly in the insulation, the secondary requiring much higher insulating properties than the primary. There are, generally speaking, two classes of ignition cable, viz., plain rubber insulated and braided cable. Rubber is a very good insulating material, but it is subject to attack by oil, and particularly by gasoline, and as it is impossible to keep these off the ignition cables on a car, the plain rubber cable "rots" in time at points where oil or gasoline may collect on it. For this reason it has become the custom to provide the cables with cotton braiding over the rubber and to impregnate the braiding with an oil-resisting enamel. The braiding serves chiefly as a web to carry the enamel. However, some cables are made with a coating of flexible enamel applied directly to the rubber and others are provided with a braiding, but are not impregnated with enamel. The rubber insulation is compounded with a very considerable proportion of mineral matter and is vulcanized. Primary cable is usually from 3/16 to ¼ inch in diameter and secondary from ⅜ to ½ inch. Primary cables are sometimes made in multiple form, two, four or six strands being combined in a single cable, which is especially adapted for connection from a multiple coil to the timer.

Cable Terminals and Connectors—The reliability of modern ignition systems is due in a considerable degree to the

CABLES, TERMINALS, SWITCHES, WIRING METHODS

substantial and positive terminals and connectors used. In early ignition practice it was customary to merely bare the end of the cable of its insulation, twist it around its binding post and screw down the nut. Not only were such connections insecure and frequently came loose, but the constant twisting back and forth of the wires every time a connection was made or broken soon caused them to break off. Cables are now always provided with terminals which insure good electrical contact and obviate the necessity of bending the wires in making and unmaking the connections. One particular type is stamped from sheet copper and is of the form shown in Fig. 157. The end of the cable is freed of its insulation and the

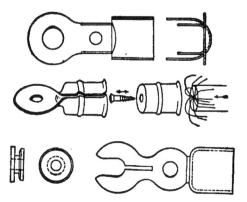

FIGS. 157, 158 AND 159—CABLE TERMINALS.

stranded copper wire is passed through the small hole in the terminal, to hold the wires in place while they are being soldered on the inside, after which the portion extending through the hole may be clipped off. The terminal is provided with two pairs of flaps of which the larger pair is closed to clasp the outside of the insulation and the smaller pair to clasp the wire strands. Terminals on low tension cables should always be soldered to insure good electrical contact. On high tension cables it is not necessary to use solder. The eye of the terminal is passed over the binding post and is secured in position by a nut which is preferably locked by a check nut. Where it seems undesirable to have a consider-

202 CABLES, TERMINALS, SWITCHES, WIRING METHODS

able length of the conductor exposed, after the terminal is secured to the cable it may be wound with adhesive tape.

Another form of terminal, the Herz, is illustrated in Fig. 158. The cable is freed of insulation for about ¾ inch, then the wires are bent back over the cable, as shown in the sketch, the copper shell is pushed over them and the brass screw inserted through the hole in the top of the shell and screwed into the strands of wire. This inner shell can be pushed into an outer shell with an eye passing over the binding post, and the terminal is said to be detachable.

A spring clip terminal which can be quickly put in place and removed is illustrated in Fig. 159. It is made with a cap to go over the insulation of the cable and the wire is passed through holes in the cap and the flat portion of the terminal and is secured either by twisting it around the terminal or soldering it to same. With the terminal comes a ferrule, as

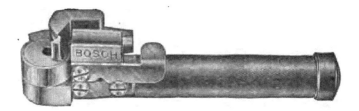

FIG. 160—BOSCH TERMINAL TOOL.

shown, which is secured to the binding post. The terminal is slotted, as shown, the smallest width of the slot being slightly less than the diameter of the throat of the ferrule, so the terminal must be forced over the latter and is held in place thereon by spring pressure. It is simply pushed on and pulled off. Such clip connectors are well suited for spark plugs, timers, etc.

Terminal Tool—The Bosch Co. has brought out a tool for forming a terminal on the cable (see Fig. 160). In the case of a cable with heavy insulation (high-tension cable), a straight brass tube or sleeve (1, Fig. 161) approximately 1⅜ inches long, of the proper diameter to fit the wire snugly, and with one end beveled to a sharp edge, is forced over the stranded wire. First, however, the insulation is stripped from the cable for a distance of approximately ⅞ of an inch, after which the terminal tube is slipped over the conductor with

CABLES, TERMINALS, SWITCHES, WIRING METHODS 203

the beveled edge first, until the end of the conductor is flush with the outer end of the terminal tube (2, Fig. 161). By means of the tool shown in Fig. 160 the whole is then bent into a loop, as shown at 4, Fig. 161. While in the process of bending the tube is somewhat flattened and firmly grips the wire strands. Besides, since the tube is forced under the insulation for a distance of about ½ inch, the tendency of the wire to break where it leaves the insulation is eliminated. In the case of cable provided with comparatively thin insulation, it is impossible to force a tube under same, and in that case the tube is expanded at one end as shown at 5, Fig. 161, thus providing a ferrule which slips over the insulation. This

FIG. 161—METHOD OF FORMING TERMINAL WITH TOOL.

gives the required rigidity at the point where the wire leaves the insulation and at the same time prevents the braided cover of the cable from unraveling.

Switches—Switches for ignition purposes are of two kinds, viz., hand switches and foot switches or kick switches. The latter, of which illustrations were given in Figs. 140 and 145, are made with a very substantial "handle" of such shape that it can readily be operated with the toe of the shoe. This type is especially convenient if the switch must be located at such a distance from the driver's seat that it cannot well be reached by the hand. Hand switches are again divided into turnover switches and pull switches. The former comprise a number of insulated contact points over which a switch lever or rotary disc is adapted to move. The contact points or buttons connect with binding posts on the back of the switch to which the cables are connected. Pull switches as a rule are used only for opening and closing a single circuit, such as the primary circuit of a magneto. One design of push and pull switch is shown in section in Fig. 162. A plunger H of insulating material extends through a central hole in the metallic switch base and carries at its inner end a guide F as well as a split pin which enters a socket in an

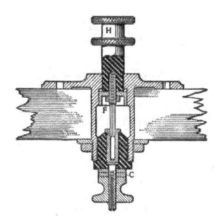

FIG. 162—PUSH-AND-PULL SWITCH.

insulated terminal C. When the plunger H is moved in or out the split pin slides in its socket, but it never completely leaves this. In the cut the switch is shown in the closed position, and the circuit is opened when the guide F leaves the guide surface of the switch base as the plunger is pushed in.

With some systems a reversing switch is used so that every time the circuit is closed the current will flow through the interrupter in the direction opposite to that in which it flowed the last time. The principle of such a switch is illustrated in Fig. 163. On a circular base there are four equally spaced contact buttons. We will say the battery terminals are connected to two buttons located directly opposite each other

CABLES, TERMINALS, SWITCHES, WIRING METHODS 205

and the coil primary terminals to the other two buttons. The movable part of the switch has two contact sectors of such dimensions that each will span adjacent contact buttons. For the position of the switch represented in the drawing the current flows in the direction indicated by the arrows, and it can readily be seen that if the switch is turned a quarter revolution so the contact sectors occupy the positions indicated in dotted lines the current will flow through coil and interrupter in the opposite direction. Midway between these two positions the switch is open.

Usually some form of safety device is incorporated in

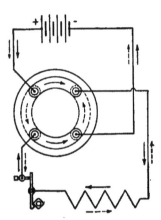

FIG. 163—ILLUSTRATING PRINCIPLE OF REVERSING SWITCH.

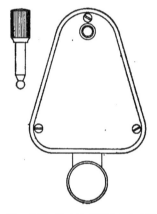

FIG. 164—LEVER SWITCH WITH REMOVABLE PLUG.

switches, affording a means of insuring that the car cannot readily be tampered with in the absence of the operator. Modern "kick" switches often are provided with a key by means of which the switch can be locked in the "off" position. A somewhat simpler plan is to use a metal plug which, when inserted into the switch, bridges a gap in the circuit. When he wants to leave his car the operator carries this plug along in his vest pocket and it is then impossible for any one not specially familiar with ignition circuits to close the circuit and start the motor. A switch with removable plug is illustrated in Fig. 164. It will be understood that the fulcrum of the switch lever is insulated and that the plug puts this fulcrum in metallic connection with the switch base, which is grounded.

Wiring Methods—In wiring the ignition system efforts must be made to arrange the wires in a neat manner, to secure them firmly so that they are not likely to become broken as a result of continuous vibration, to protect them as much as possible against the likelihood of mechanical injury while work is being done around the engine, and—in the case of the sec-

FIG. 165—CONNECTIONS FROM MAGNETO ON ONE SIDE OF ENGINE TO PLUGS ON THE OTHER.

ondary cables—to prevent interference with the proper operation of the system by electrostatic induction.

Most importance, of course, attaches to the arrangement of the cables near the motor, as these are in plain view when the bonnet is lifted and are most likely to interfere with access to important parts if injudiciously placed. In low-priced four-cylinder cars where the high tension distributer is placed at the side of the cylinder block about midway of its length, it is customary to run the high tension cables from the distributer to the spark plugs each in a more or less direct line, supporting the cables only by a bracket or cleat at the top of the engine. If the plugs are located on the opposite side from the distributer the support is often made integral with the water return manifold. An arrangement employing cleat supports is shown in Fig. 165.

In some of the battery systems the combined timer and distributer is carried on top of a vertical shaft at the height of the cylinder heads and midway between the front and rear of the engine, so the cables from it to the spark plugs, even in the case of a six-cylinder motor, do not need a special support.

A neater effect is produced, especially if the distributer is located at one end of the engine, by running the cables through

FIG. 166—CARRYING HIGH TENSION CABLES IN A METAL OR FIBER TUBE.

a tube running lengthwise of the engine above the cylinder heads (see Fig. 166). There are outlets in this tube either opposite each spark plug, as in the illustration, or between pairs of plugs, in which case two cables leave the tube together. The tube may be either of brass or fiber, but a brass tube seems to be used most extensively.

Electrostatic Interference—It was pointed out above that the high tension cables must be so arranged as to obviate interference due to electrostatic induction between adjacent ones. In order to explain the latter phenomenon it will be necessary to state some of the principles governing static electricity.

Every conductor has a certain capacity for absorbing an electric charge. If the conductor stands alone at a considerable distance from any other conductor its capacity depends merely upon its dimensions and form. But if we bring another conductor near it, its capacity increases. The capacity in the case of such opposing conductors (condenser) also depends upon the specific inductive capacity of the intervening medium, which, in the case of a cable, would be made up partly of its insulating material and partly of air.

Every electric conductor in the neutral state contains equal charges of positive and negative electricity. In all discussions of electrostatic phenomena, the earth is supposed to be neutral or of zero potential. Positive electricity is that kind which is generated on a glass rod when it is rubbed with a woolen cloth, and negative electricity that kind generated on a piece of resin if similarly rubbed. The glass rod, then, has a higher potential than the earth and the resin a lower potential. Electric charges of the same sign repel each other and charges of opposite sign attract each other.

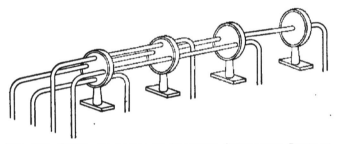

FIG. 167—CABLES SUPPORTED BY BRACKETS CONTAINING CIRCULAR DISCS OF INSULATING MATERIAL.

Now let us assume two high tension cables lying parallel to each other and close together. One of these is connected through the high tension distributer to the source of the high tension current, the secondary winding of the coil or armature, and both, of course, are connected to spark plugs. One end of the secondary winding in which the high tension impulse is induced is connected to ground. As the electromotive force in the secondary winding of the coil or armature rises, a charge will flow into the cable to which the winding is connected. This charged conductor then acts by electrostatic induction on all other conductors near it, such as the secondary cable above referred to. If the core of the latter cable were connected to ground, a charge of the same sign as that on the live cable would flow from it into ground. Since the cable is insulated, no charge can leave it, but it is

CABLES, TERMINALS, SWITCHES, WIRING METHODS 209

obvious that the electrostatic induction takes place just the same and that there is a tendency for a charge to pass from the insulated cable to ground. This tendency may under certain conditions become so strong that the resistance of the air gap of the spark plug to which this cable is connected (and which gap forms the shortest path to ground) is broken down and a spark occurs at this plug, which is out of all metallic connection with the source of secondary current. This effect is known as electrostatic interference.

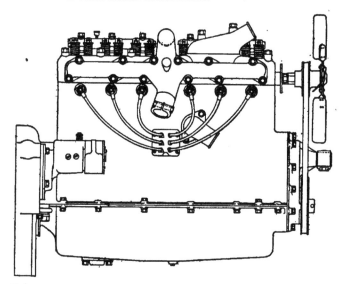

FIG. 168—ON THE GRANT SIX-CYLINDER ENGINE THE HIGH TENSION CABLES ARE CARRIED THROUGH A CHANNEL IN THE CYLINDER BLOCK.

As pointed out, the particular cable which carries the current at any moment, together with the other cables of the high tension system and any other nearby metal parts, forms a condenser. This is subject to the law of the condenser, viz.,

$$\text{Electromotive force} = \frac{\text{Quantity of charge}}{\text{Capacity}}$$

Now we have seen that the quantity of electricity which a coil system will throw into the circuit on the interrupter being opened is limited to the equivalent of the magnetic energy stored up in the coil core. Therefore, if the capacity of the cable is high, the electromotive force active in it may never attain the value necessary to produce a spark.

To avoid misunderstanding, it should be pointed out that both of these phenomena, viz., a spark at the plug not included in the high tension circuit at the moment, due to electrostatic induction between adjacent cables, and failure to produce a spark because of excessive capacity of the spark plug cable, are rather remote possibilities. They are known, however, to have occurred in cases where the high tension cables were exceedingly long and run very close together. Where the cables of necessity must be long it is well to observe certain precautions. If each cable were encased in a grounded metal tube or sheath, there would be absolutely no electrostatic induction between cables. This, however, is not very practical as such a metal casing would greatly increase the capacity of the cables. One method suggested is to support the cables by grounded metal clips short distances apart. This keeps them apart and also reduces the electrostatic induction in another way, as the clips, applied in the way shown, form a sort of screen between adjacent cables, their screening effect being probably increased by a slightly conducting film of dirt or moisture which will collect on the outside of the cable insulation. This method of arranging the cables is not always applicable, because it requires considerable space.

Another method recommended is that illustrated in Fig. 168, where the cables pass through circular blocks of insulating material carried in metal fittings screwed into the cylinder heads. This is a symmetrical arrangement and keeps the individual cables the requisite distance apart to prevent electrostatic interference. A very neat arrangement of the cables is found on the Locomobile six-cylinder motors. They are carried in a wooden container running along the top of the cylinders, each cable lying in a separate groove. The container is not far from the spark plugs and the lengths of cable from the container to the plugs are very short.

CHAPTER XX

Storage Battery Charging—Magnet Recharging

Where storage batteries are used for ignition on a car not equipped with an electric generator, they must be recharged periodically. The current required for this purpose may be obtained either from electric service mains or from an electric generator specially installed for the purpose. In the great majority of cases the current is taken from service mains. These carry either direct or alternating current, generally the latter. Only direct current can be used for charging, and where the service is alternating the current must be rectified before it can be used. Aside from the fact that the current must be direct or continuous, it must also be of the proper voltage. A voltage of 110-115 has become practically standard in electric lighting practice, and with this voltage batteries of 42-44 lead cells in series can be charged to the best advantage. For each cell in series, therefore, there is required a charging voltage of at least 2.65 volts.

Charging from Direct Current Mains—The simplest method of charging an ignition battery from direct current lighting mains is by means of a so-called current tap or charging plug which screws into an ordinary lamp socket (Fig. 169). The tap is provided with, say, three sockets into which lamps

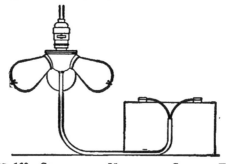

FIG. 169—CHARGING BY MEANS OF A CURRENT TAP.

may be screwed, and with a charging cable which connects to the battery to be charged. The three lamps are connected in parallel and the battery is in series with all of them, so that it receives as much current as the three lamps together. As a rule such current taps do not take more than three lamps, and the charging rate is limited by this.

A more substantial installation for charging a battery from direct current mains is shown in Fig. 170. From the two mains, wires of the grade usually employed for house wiring are run to a small switchboard. There a 10-ampere fuse is inserted in each line and below the fuses is placed a single throw double pole switch to which the wires are connected. To one of the lower terminals of the switch are connected six or seven 100-watt lamps, as shown, and between the other lower switch terminal and the lamps is connected the battery.

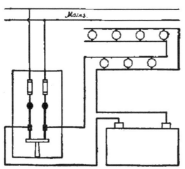

FIG. 170—CHARGING BATTERY FROM DIRECT CURRENT MAINS.

Determining Polarity of Mains — In making the connections, care must be taken that the positive terminal of the battery connects with the positive main, as otherwise the battery would become discharged and reversed. A wrong connection will be shown instantly when the switch is closed, by the lamps lighting up more brightly than normally, while they will burn more dimly than normally if the connections are correctly made. It is, however, preferable to determine the polarity of the charging mains in advance. This can be done by inserting the two wires to be connected to the battery into a glass of slightly aciduated water and turning on the current. Gas bubbles will then be observed forming at both electrodes and rising to the surface. The rate of gas development is far greater at the negative than at the positive terminal, which can thus be identified. Another method of determining the polarity of charging mains consists in moistening a strip of red litmus paper, placing the ends of the two wires upon this paper a short distance apart and then turning on the current. A blue spot will be formed under one of the two wires—the negative one.

Economics of Battery Charging—As regards the lamps, it is most economical to use carbon filament 32 candlepower lamps which consume about 100 watts each. Twice as many 16-candlepower carbon filament lamps would be needed for the same charging rate, and tungsten lamps cost more and have a shorter life than carbon lamps. Since the battery requires about 8 volts to force a current through it, in the case of a 110-volt circuit there remains only 102 volts for the lamps and these will burn dimly. A 100-watt lamp on 110 volts consumes 10/11 ampere, but on 102 volts the current consumption is less, something like 6/7 ampere, so that the seven lamps in parallel will allow about 6 amperes to pass. It will be understood that in this case only about 8 volts are utilized for charging and 102 volts are wasted in the lamps, so that the method is really very inefficient. If 100 ampere-hours had to be put into the battery the total energy consumption would be

$100 \times 110 = 11000$ watt-hours or 11 kilowatt-hours,

and at 10 cents per kilowatt-hour the cost for current figures out to $1.10, while the cost of that portion of the electric energy actually absorbed by the battery is only 8 cents. If two or three 6-volt batteries can be charged at the same time they may be connected in series. This will slightly reduce the rate of charge, but the cost of current for charging the two or three will be no greater than that for charging a single battery.

Commercial Installations—While charging from service mains through lamp resistances is expensive as regards the cost of electric energy consumed, it is very convenient, as the current is self-regulating and practically no attention need be paid to the batteries while on charge. Therefore, in small garages where a moderate amount of charging business is done, a system similar to that described can be used to advantage. From three to five batteries should be connected in series so that at least one-quarter of the energy consumed will be stored in the batteries. The lamps may be arranged in the form of a rack high up on the wall, say above the switchboard, and the batteries may be placed on shelves on the wall directly above which are binding posts to which the battery terminals are connected.

Whenever batteries are recharged their electrolyte should be replenished, and to make this more convenient it is well to place on an upper shelf bottles of distilled water, acid and electrolyte. These three bottles should be provided with glass tubes passing through their stoppers to near the bottom, rubber tubes being slipped over the outer ends of the glass tubes, of such a length that they will reach to the most distant battery. By means of these tubes either acid, acid solution or distilled water—according to the requirements as shown by hydrometer test—can be siphoned into any cell, the siphonic action being controlled by pinch cocks on the tubes.

Storage battery electrolyte is very corrosive and the fumes rising from the battery when the latter is gassing toward the end of a charge tend to attack exposed copper terminals. All battery connections are generally made of lead and are protected by a coating of insulating paint. Any bright metal parts near the battery may be protected from the corrosive action of the electrolyte by giving them a coating of vaseline.

When more than four or five batteries have to be charged at a time the lamp racks and battery shelves may be duplicated. But in the case of a very large installation it would be more economical to transform the current from 110 volts to a lower voltage and use some form of low resistance rheostat in place of the lamps, so as to cut down the waste in the regulating resistance. For transforming the current, use is made of a rotary converter which consists of a dynamo with two armature windings and two commutators. Current from the mains is sent through one winding and charging current is collected from the other, the current being multiplied and the voltage reduced in substantially the ratio of motor to generator armature turns. When charging through lamps the charging current can be calculated from the number and size of lamps in circuit; in all other cases it is advisable to include an ammeter in the circuit by which the charging current may be regulated.

Charging from Alternating Current Circuits—If only alternating current is available the first thing required is a rectifier by means of which the current can be changed to a direct current. There are five different types of rectifiers, viz.:

  The electrolytic rectifier,
  The vibrator type rectifier,
  The mercury arc rectifier,
  The Tungar rectifier, and
  The motor generator.

As in the case of charging from direct current circuits, the voltage also must be reduced to suit the requirements of the battery, and in the case of an alternating current this can readily be accomplished by means of a transformer, as illustrated in Fig. 46, which contains no moving parts and therefore requires no attention.

Electrolytic Rectifier—The electrolytic type of rectifier is

STORAGE BATTERY CHARGING

based upon the principle that an electrolytic cell containing one aluminum and one other electrode will allow a current to pass only in one direction. In practice one of the electrodes is made either of aluminum or of an aluminum-zinc alloy, and the other electrode of iron. The cell is filled with saturated solution of ammonium phosphate. Suppose such a cell to be connected to alternating current mains. When the current passes in the direction from the aluminum to the iron electrode, a thin layer of oxides and phosphates of aluminum is almost instantly formed on the aluminum electrode, and this layer offers such a high electrical resistance that it practically cuts off the current until the direction of the latter is reversed. The reverse current instantly reduces the high resistance compounds forming the layer and then flows freely through the cell.

With a single cell only one wave of the current can be used, the other wave being stopped, but by using four cells connected as shown in Fig. 171, both waves of the alternating current can be used for charging. It will be seen from the illustration that the four cells are connected in two series, and the two cells in each series are connected in opposition, so that no current can flow through either series in either direction. The battery to be charged is connected between the two cells of either series, and this gives a free passage across the line for both impulses of a period. The positive impulse will flow, say, through the upper right-hand cell, the battery and the lower left-hand cell, while the negative impulse will flow through the lower right-hand cell, the battery and the upper left-hand cell.

The efficiency of the electrolytic rectifier is only about 70-75 per cent, and the cell heats up rather quickly. For this reason some electrolytic rectifiers are provided with radiators. Fig. 172 shows a charging stand comprising an electrolytic rectifier for charging ignition and similar batteries. All of the apparatus is mounted on an angle iron frame. On top there is a switchboard containing an ammeter, a main line switch, a charging circuit switch and a regulating switch. Below this switchboard is located the transformer for reducing the voltage. One winding of this transformer is wound in sections and leads are brought out to the regulating switch so that the number of effective turns in this winding can be varied. At the bottom is seen the rectifier with radiator. Such a rectifier gives a pulsating current, but this is just as good for battery charging as a continuous current.

Vibrator Type of Rectifier—This type is based upon the action of a vibrator operating in synchronism with the alternating current. Fig. 173 is a diagram of a rectifier of this type made by the Westinghouse Co. Alternating current of, say, 110 volts enters at the terminals AA and passes through the double pole switch B to the primary coil C of the transformer. The secondary or low voltage winding of this transformer is divided into halves, each of which is connected through a regulating resistance E with the stationary contacts I and J of the vibrator. Opposite these contacts is located the vibrating magnet O, which carries coils HH connected to the battery charging terminals LL. As long as the battery is in circuit, magnet O is energized. Directly above magnet O are located two stationary electromagnets F and G, whose windings are connected in series and supplied with current from one-half of the secondary winding of the transformer. These two stationary magnets are so connected up that their lower ends are always of like polarity. Magnet O is pivoted at its center, and is held by a couple of flat springs, not shown in the diagram, in a neutral position, when both contacts I and J are open. The polarity of magnets F and G, which are energized by an alternating current, changes at the same rate as the direction of the alternating current.

As long as the battery is con-

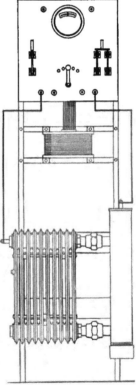

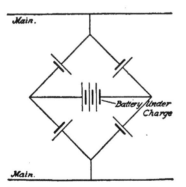

Fig. 171—Arrangement of Four Electrolytic Rectifier Cells.

Fig. 172 — Electrolytic Rectifier Charging Set.

STORAGE BATTERY CHARGING 217

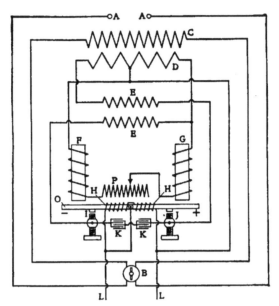

Fig. 173—Diagram of Vibrator Type Rectifier (Westinghouse).

nected in circuit, magnet O is energized and its polarity never changes. Assume that the current flows around it in such a direction as to give it the polarity indicated in the diagram. Also that at any particular instant the lower poles of magnets F and G are positive, then the negative pole of magnet O will be attracted by magnet F and magnet O will establish contact with stationary contact J. Current will then flow from the left-hand end of secondary winding D through the upper regulating resistance E, through contact J into magnet O and out therefrom through the left-hand binding post L, through the battery and thence through the other binding post L back to the center point of secondary winding D. At the end of a half cycle the circuit is broken at J and magnet O returns to its normal position. Then the current through magnets F and G reverses and the lower poles become negative. Now the positive end of magnet O is attracted by magnet G and magnet O makes contact with stationary contact I. The current now flows in the opposite direction through the secondary winding S and it passes from the right-hand end thereof through the lower regulating coil E, through contact I, magnet O, into the battery and through left-hand terminal L, in the same direction as previously. This keeps up as long as the battery is connected to the rectifier. Should the source of current fail for any reason, a pair of springs acting on magnet O will hold it in a neutral position in which both contacts remain open, so that the battery cannot discharge. A pair of condensers is shunted across the two vibrators respectively to prevent arcing.

Mercury Arc Rectifier—This rectifier is in a way similar in principle to the electrolytic rectifier, in that it constitutes a sort of electrical check valve which permits current to flow through in one direction but not in the other. It consists essentially of a glass bulb from which the air has been exhausted, containing a pool of mercury in the bottom, with a negative terminal or cathode fused into the glass and extending into the mercury. Two positive terminals or anodes pass through the upper portion of the wall of the bulb, and in addition there is an auxiliary anode used only for starting, which is located close to the surface of the mercury pool. Current will flow from an anode to the mercury, but will not flow from any point to an anode or positive terminal Operation does not begin, however, until an arc is formed at the surface of the mercury, with the mercury as the negative terminal. The mercury then begins to fill the globe with vapor which forms, in fact, an arc between the cathode and the two anodes in turn. Starting is effected by slightly tilting the bulb so that the starting electrode touches the mercury.

The general arrangement of a mercury arc rectifier battery charging installation is shown in Fig. 175. Alternating current from the mains is sent through a transformer to reduce its voltage. The secondary or low tension winding is divided into halves. Its two outer ends are connected to the anodes of the rectifier, and its central terminal and the cathode of the rectifier are connected to the battery to be charged. Then the electromotive force induced in that half of the secondary winding which at the moment is the negative end will send a current through the battery and the rectifier. No current will flow in the other end of the transformer, owing to the check valve action of the rectifier. The next moment the electromotive force in the secondary winding reverses and the other half of this winding now sends a current through the battery and rectifier. The current given out by such a rectifier is a direct pulsating current which drops to zero between succeeding pulsations. By including a choke coil in the charging circuit it is possible to prevent the current from ceasing entirely, as the choke coil absorbs energy during the increase in current and gives it out again during the decrease, thereby slightly extending the time of each pulsation, so that succeeding pulsations overlap. The efficiency of a mercury arc rectifier for charging low voltage batteries is limited by the fact that there is a constant voltage drop of about 14 volts in the rectifier.

The Tungar Rectifier—A new form of rectifier has been de-

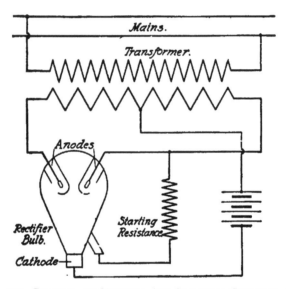

Fig. 174—Diagram of Mercury Arc Rectifier Charging Set.

veloped by the Research Department of the General Electric Co. and marketed by the company as the Tungar. It is based on the principle that a vacuum bulb with one hot and one cold electrode acts as a rectifier. During that part of the cycle when the hot electrode is negative, electrons or small particles of negative electricity are projected from it toward the cold positive electrode, these electrons forming current carriers. There is no corresponding action during the other half of the cycle when the hot electrode is positive. The principle was not directly adaptable to commercial rectifiers, as in vacuum tubes of this type, known as Kenotrons, there is a voltage drop of from 100 to 500 volts between anode and cathode when the tube is in operation. The bulbs are therefore filled at low pressure with argon, an inert gas found in the atmosphere. When the heated electrode, which is in the form of a closely coiled .tungsten wire filament, is negative, it emits electrons which collide with the molecules of the argon and ionize them, i.e., render them conductive in the direction from the positive electrode or anode, which is in the form of a fairly large block of graphite, to the negative electrode or cathode, the filament. During the other half cycle, when the filament is positive, any electrons that may be emitted are driven back to the electrode and the argon remains non-conductive. Impurities in the argon tend to cause disintegration of the cathode (filament). Therefore, although the argon is introduced into the bulb—after the latter has been completely exhausted of air—in the purest possible form, a purifying agent has to be put into the bulb to keep it pure. This purifying agent is in the form of a wire ring on the anode.

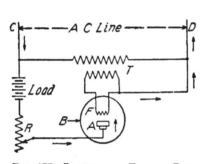

FIG. 175—DIAGRAM OF TUNGAR RECTIFIER.

In Fig. 175 are shown the connections of the simplest form of Tungar rectifier, which utilizes only a half wave of the alternating current. The equipment consists of the bulb B with filament (cathode) F and anode A, transformer T for exciting the filament, rheostat R and the load, which latter is shown as a storage battery.

At an instant when side C of the alternating current supply is positive, the current flows as indicated by arrows through the storage battery, rheostat, bulb, and back to the opposite side of the alternating current line. A certain amount of the alternating current, of course, flows through transformer T to excite the filament, the amount depending on the capacity of the bulb. When the alternating current reverses and the side D becomes positive, the current is prevented from flowing. In other words, the current is allowed to flow from the

STORAGE BATTERY CHARGING

anode to the cathode or against the flow of emitted electrons from the cathode, but it cannot flow from the cathode to the anode with the electrons.

The smallest size of Tungar rectifier made has a capacity of 2 amperes on a three-cell battery and 1 ampere on a six-cell battery. It comprises a bulb and receptacle, a compensator and reactance on one core and a three-ampere fuse and receptacle. The bulb is about 2 inches in diameter and between 5 and 6 inches long. The compensator is designed to reduce the primary voltage from 115 to that required to deliver the proper direct current voltage. The connections of the compensator and rectifier complete are shown in Fig. 176. This device weighs 8 pounds.

In Fig. 177 is shown a somewhat larger rectifier of the same type, which will charge either three or six cells of lead battery at 6 amperes from a 115-volt circuit. Its construction is the same as that of the 2-ampere type described except that there are taps for charging either three-cell or six-cell batteries. The voltage drop in the Tungar bulb is from 5 to 10 volts and the efficiency of operation of the 6-ampere rectifier at full load is 75 per cent.

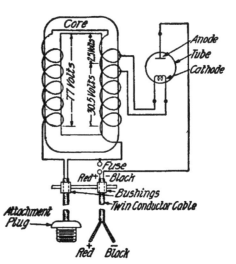

FIG. 176—CONNECTIONS OF TUNGAR RECTIFIER WITH COMPENSATOR.

Motor Generator—A motor generator, as the term implies, combines a motor and generator, the motor in this case being alternating and the generator direct. By giving the generator a suitable winding the machine can be made not only to change the character of the current from alternating to direct, but also to reduce (or increase) the voltage in any desired

222   STORAGE BATTERY CHARGING

ratio. The voltage of the direct current, and hence the rate of charging, can be controlled by means of a rheostat in the generator field circuit, which involves a minimum loss of energy. This same method of control is applied if a battery is charged from a special low voltage generator driven by belt. Fig. 177 shows a battery charging motor generator manufactured by the Westinghouse Electric & Mfg. Co. It is designed to operate on 110-volt, 60-cycle alternating current circuits and will charge either one six-volt battery, one 12-volt battery, or two six-volt batteries in series, control being by field rheostat. The output is 10 amperes.

FIG. 177—SIX AMPERE TUNGAR RECTIFIER

Recharging Magneto Magnets—The permanent magnets of ignition magnetos in the course of time become weakened, with the result that the engine will misfire at low speeds. Inasmuch as misfiring at low speed may be due to other causes than weakness in the magnets, it is best, if the latter trouble is suspected, to determine the length of spark the magneto will develop in the atmosphere. If it does not produce a spark at least ¼ inch in length it is a sign that the magnets are weak, and it is then advisable to remagnetize or recharge them. Generally the magnets are taken from the magneto. A recharger in the form of an electromagnet, to which the magnets can be applied, may be used. This can be energized either from direct current service mains or from a 6-volt storage or dry cell battery. A design for a magnet recharger is shown in Fig. 179. The limbs of the magnet are made of soft steel 1 inch in diameter and 3 inches long. They are secured to a base measuring 5¼ x 1½ x ⅝ inches and are provided with pole pieces measuring 1¾ x 1¾ x ⅝ inch. All contacting surfaces should be machined absolutely flat and square so that there will be good metallic contact over the entire surfaces. Before the wire is wound on the magnets the latter must be insulated. A sort of spool is formed on them by means of two fiber rings, and in this connection it is best if the magnet cores are turned down from a diameter of 1⅛ inches and a thin collar is left on them at one end which supports the fiber ring at that end, the other ring being supported by the pole piece. The core between the fiber rings

FIG. 178—MOTOR-GENERATOR (WESTINGHOUSE).

is then wrapped with several layers of muslin which is given a coat of shellac in alcohol and allowed to dry.

The winding to be applied depends upon the voltage of the source of current to be used. For a six-volt battery wind on three layers of No. 12 double cotton covered magnet wire; for a 110-volt circuit, eight layers of No. 22 double cotton covered magnet wire. The ends or leads of the wire are taped and the outsides of the coils shellaced to make their exposed cotton insulation more enduring. Mount the whole on a wooden base which also carries a single pole single throw knife switch and a binding post. Connect the two coils together so that if the current flows through one right handedly it flows through the other left handedly (both looked at from the top). Connect one free end of the coil to one terminal of the switch and the other to the bottom of the binding post on the base. The current source is connected to the other terminal of the switch and to the binding post. (See Fig. 180.)

In recharging the magnets it is important that they be applied to the recharger with unlike poles together, that is, north pole of magnet to south pole of recharger and vice versa. Since like poles repel and unlike poles attract each other, the magnet finds its own position if freely held a short distance above the recharger poles, while the current is switched on.

In recharging, set the magnet on top of the charger and switch on the current, rock the magnet back and forth on its pole edges a number of times, then lay it on its side with the poles away from you and extending just beyond the far

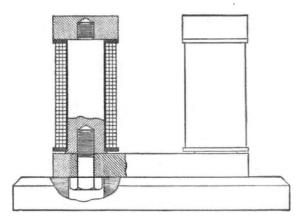

FIG. 179—MAGNET RECHARGER.

edges of the recharger poles, apply a keeper to the pole pieces, switch off the current and withdraw the magnet sideways from the recharger. The keeper should remain in place till the magnets are again applied to the poles of the magneto.

The windings specified above will heat up quickly when connected to current sources of the voltage mentioned, and the switch should never be left closed for more than a few minutes at a time.

Where direct current mains are accessible the magnets may be recharged without demounting them. The wire must then be wound directly over the magnets, and lamp cord is probably best for the purpose, as it is very flexible and well insulated. The bared ends of the cord should be twisted together so that the two strands form one conductor. Wrap on about 50 turns and connect the wires to the main switch of the installation through a 10-ampere fuse. A very intense current will flow through the wire for an instant, until it is stopped by the blowing of the fuse. Particular care should

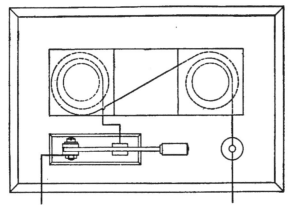

FIG. 180—CONNECTIONS OF MAGNET RECHARGER.

be exercised to so make the connections that the magnets will not have their polarity reversed, for with a given magnetizing force the remanent magnetism will be greater if the polarity is not reversed. This method obviates the necessity of taking the magneto apart and undoubtedly involves the least amount of labor.

CHAPTER XXI

Ignition Testing Apparatus and Tests

In the ordinary use and care of ignition apparatus it is seldom that any testing instruments are required except a voltmeter and an ammeter. But in shops making a specialty of testing and repairing ignition apparatus, and especially in the experimental and testing departments of manufacturers of such apparatus, various testing instruments designed to indicate faults or to indicate or record the exact performance of the apparatus can be used to advantage.

Fault Finders—There have been a few devices designed to enable operators to locate ignition faults, but they have not come into extensive use. One such device (Fig. 181) is an air-tight chamber made partly of glass and partly of metal, with an opening into which a spark plug can be screwed and another opening through which air pressure can be pumped up in the chamber. This device is particularly suited for tests with multiple coil systems to determine whether a miss is

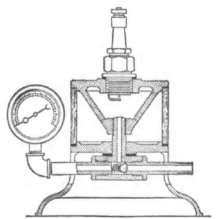

FIG. 181—SPARK PLUG TESTER.

due to defects in the plug, in the coil or in the connections.

Suppose one of the plugs in the engine is missing. Ordinarily a spark plug known to be all right is in position in the testing chamber. The cable leading to the spark plug in the cylinder which is missing fire may then be disconnected and connected instead to the spark plug in the testing chamber. The current is then turned on and the result noted. If the spark jumps the gap at atmospheric pressure, air is pumped into the chamber until the pressure equals the working compression of the engine. If the coil is perfect, there is a flash when the current is turned on. If not, it indicates that the coil has broken down or is short-circuited. The spark plug and wiring may be tested in a similar manner, or, if desired, the spark plug thought to be defective may be removed from the cylinder and screwed into the opening of the testing chamber. By then connecting the cable to the spark plug the test can be made at both atmospheric and engine pressures.

In another testing device for ignition systems, a set of spark gaps about ¼ inch long in the atmosphere and a corresponding set of vacuum tubes are combined in a box similar in appearance to a photographic camera. If the spark cable carries a normal voltage, a spark will jump across the spark gap of the tester; if the voltage is much below normal, no spark will pass, but the vacuum tube will produce the well-known fluorescence, and if the cable is absolutely dead there will be neither a spark nor fluorescence. From the results obtained certain conclusions as to the location of the trouble may be drawn.

Cable Testing Device—All ignition cable is guaranteed by the manufacturers to withstand certain insulation tests, and it is therefore necessary to test the cable before it leaves the factory. In Fig. 182 is shown an apparatus for making such tests, which is the invention of N. A. Wolcott, of the Packard Electric Co. The cable tester comprises an induction coil capable of producing a 3-inch spark in the atmosphere. The primary coil P is connected through switch D to a storage battery B. Secondary winding S is connected to adjustable spark gap G and also to terminals EF, which are metallically connected to contacts HJ. The latter are spring contacts which brush the exterior of the cable as the wire is drawn through. K is a guide, which, together with contacts H and J, is mounted on a heavy wooden base firmly bolted to a support.

The cable A is drawn rapidly through the guide K and contacts E and F, while the spark is continuously jumping spark gap G. If a defective spot in the insulation passes through the tester, it is immediately made apparent by the absence of sparking at G and vigorous sparking first at F and then at E as the bad spot passes these respective points. The sound of the sparking at F and E is entirely different from that at G, which affords additional means of detecting the defect if the operator happens not to be looking at the tester the instant the sparking occurs.

Butler Magneto Tester—A device for determining the maximum length of spark which a magneto will give in the atmosphere at different speeds of rotation, the maximum timing range, the effective timing range and the volume of the spark, has been developed by the Butler Mfg. Co. of Carthage, Ind. It consists of a cast aluminum base having a headstock A at one end and an adjustable table B for supporting the

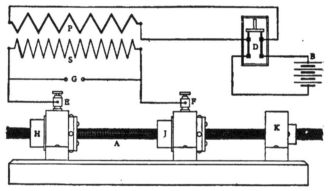

Fig. 182—Wolcott's Cable Tester.

magneto at the other. The magneto is fastened to the table by means of a chain clamping device, as shown in Fig. 183, and the magneto armature shaft is connected to the headstock spindle by means of a chuck C. A transverse shaft is driven from the headstock spindle through a pair of helical gears. This shaft at one end is provided with a disc of insulating material which carries at its center a metal button insulated from the shaft. Connected to the metal button is a radial arm D, forming an electrode, whose free end comes close to the graduated disc or protractor E carried upon a hub formed on the headstock and free to be rocked around its axis by means of a radial handle secured to it. A high tension cable is led from the high tension terminal of the magneto to a connector F on the base, and thence a spark gap finger G provided with a ball grip H of insulating material leads to the metal button carried by the insulating disc on the transverse shaft. (In the illustration the disc and button are hidden by ball grip H.) By means of the ball grip the distance between the spark gap finger G and the metal button can be varied at will. It will thus be understood that there are two spark gaps in series, one between spark gap finger G and the metal button of electrode D and the other between electrode D and disc E, which latter is grounded.

In making a test the magneto armature shaft is first clamped in the chuck on the headstock spindle and the spark gap finger G is connected by cable with the high tension terminal of the magneto. If now the crank handle of the headstock is turned, a spark will jump between the spark finger G and

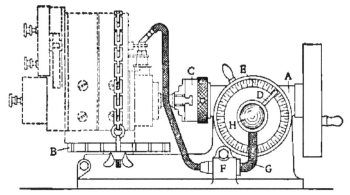

FIG. 183—BUTLER MAGNETO TESTER.

the metal button and between electrode D and graduated disc E. The electromotive force behind the spark will be shown by the distance which the spark will jump between the spark gap finger and the button, this distance being gradually increased by means of the insulated ball. The volume of the spark is shown by the distance the arc follows electrode D on the graduated disc E. It is said that with some magnetos the spark will extend over an arc of 30 degrees, while with others it will extend over less than one degree. This test for spark volume is very convenient in comparing different breaker cams, as one cam will sometimes give a spark extending over an arc of several degrees more than another.

The range of advance and retard is measured by first placing the breaker housing in full retard position, operating the crank handle, rotating the graduated disc until the sparks appear at zero, directly under the handle. Then the interrupter housing is shifted to the full advance position. Note is then taken at the point where the sparks appear and the reading of the protractor at this point gives the timing range in degrees.

To find the effective range, the interrupter housing is set in a central position between full advance and full retard; the distance between the spark gap finger G and button is increased as much as possible and still have a good, strong spark from the magneto, which should be from $\frac{1}{8}$ to $\frac{1}{4}$ inch, according to the type of magneto and speed.

After the most effective distance has been determined the protractor is adjusted until the spark appears at zero. Then the interrupter housing is shifted in a retard direction until the spark begins to weaken and note is taken of where the spark appears on the protractor, then the same procedure is gone through with for the advance direction of the interrupter housing. This will give the most effective range.

In testing a magneto and coil system with this device the cable is connected to the high tension terminal of the coil. It is also possible to test a low tension make-and-break system with this device. A battery and a non-vibrating coil are then connected in series across the spark terminals, and the high tension winding of the non-vibrator coil is connected to the spark gap finger. In testing wiring, the cable clip may be attached to various points of the wiring, distributer or to the spark plug terminal of the different cylinders of the engine. Unless the tester rests on some metal part of the engine, it is necessary to connect a wire from the ground terminal on the magneto or from the tester base to the engine.

Recorder of Synchronism—An instrument for determining whether the sparks in a multi-cylinder motor occur at exactly equal time intervals has been invented by Francis R. Hoyt, and is known as the Hitenagraph. The magneto under test is connected to the Hitenagraph by a chuck E and the driving wheel C is turned, causing the magneto A to generate current. One side of the high tension circuit is grounded. The other side connects through the high tension terminal G of the magneto and wire H to the spark index B, where the circuit is completed except for an air gap between the point of spark index B and the grounded chart drum D, so that, as the magneto is rotated, the spark in jumping from the point of B to D passes through the chart I, which has previously been placed on the drum D. F is a thumb nut which engages the index B with a worm, causing it to travel laterally across the chart,

IGNITION TESTING APPARATUS AND TESTS

thus making the sparks fall side by side. This chart is composed of blue paper, to the exterior surface of which a compound assembling paraffine is applied, so that the spark in passing through the chart melts the compound, thus leaving part of the blue paper underneath exposed. It may readily be understood that the size of this blue mark depends upon the amount of compound melted, which, in turn, is in direct proportion to the intensity or heat of the spark being generated.

The chart I travels 360 degrees while the magneto shaft moves 180 degrees, so that successive sparks from the magneto will fall side by side when the index B is moving, and

FIG. 184—THE HITENAGRAPH.

by graduating this chart in degrees one can readily see whether a magneto is absolutely synchronized, and, if not, how many degrees it is out. If the marks or indices are all exactly in a straight line, there is absolute synchronism of ignition, whereas, if there are two lines of indices, each line including every second index, then the sparks do not occur in all of the cylinders at the same period in the cycle of the engine, and the interrupter cams are not properly spaced.

Oscillographs—The current impulses in both primary and secondary circuits, which are of extremely short duration, may be studied by means of an oscillograph. Ordinary electric measuring instruments cannot be used for this purpose, because the moving parts of such instruments have too much inertia to follow such rapid variations of electromotive force and current as occur in ignition circuits. One of the most practical forms of oscillograph is that known as the iron ribbon type. The principle of this instrument is essentially the same as that of a well-known type of galvanometer. Only in this case the magnetic field is made as intense as possible

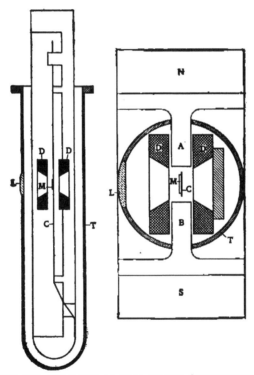

FIG. 185—DIAGRAMS OF IRON RIBBON OSCILLOGRAPH.

and the inertia of the moving parts is reduced to a minimum.

In Fig. 185 are shown diagrams illustrating the principle of this oscillograph. N and S are the poles of a powerful permanent magnet and A and B are pole shoes secured to same and designed to produce a very intense magnetic field where they approach closest to each other. A tightly stretched ribbon C of soft iron extends between the pole shoes and is enclosed

IGNITION TESTING APPARATUS AND TESTS   233

in a brass tube T. Secured to the iron ribbon where it passes the pole tips is a small mirror M and on opposite sides of the pole tips are arranged coils D through which flows the alternating or pulsating current whose wave form is to be determined. Directly in front of the mirror in the brass tube is a lense L and the coils are so shaped as to not intercept a beam of light reflected by the mirror.

When the instrument is at rest the soft iron ribbon extends with its flat sides squarely across the magnet poles, being held in this position by both the directing power of the magnetic flux and by the torsional elasticity of the ribbon. When a current flows through the coils it sets up a magnetic field at right angles to that of the magnet and the two fields together

FIG. 186—OSCILLOGRAPH (GENERAL ELECTRIC).

form a resultant. The soft iron ribbon tends to twist around until its central portion lies parallel with the lines of force, and it assumes a position which is determined by the strength of the current on the one hand and its resistance to torsion on the other. The stronger the current the more the ribbon (and with it the mirror) will deflect.

Owing to the exceedingly low moment of inertia of the ribbon it can follow the most rapid variations in current without appreciable lag.

Photographic Record of Waves—If, while an alternating or pulsating current flows through the coils, a beam of light is thrown on to the mirror, the reflected beam will constantly move up and down, and if a sensitive film were moved at uniform speed past the mirror in a direction at right angles to the plane in which the beam of light plays, then the ray of light will trace on the film a curve which represents the fluctuations of the current. In the actual oscillograph a photographic apparatus is built together with the galvanometer portion into a single instrument.

A photograph of the current wave is known as an oscillogram. A set of oscillograms were shown in Fig. 77.

CHAPTER XXI

Electric Generators

Historical—In the early days of the automobile industry, when there was keen competition between the electric, steam and gasoline propulsion systems, the inability of the gasoline motor to start under its own power stood out as one of its weakest points. Both the electric and the steam motor start from rest with a torque several times their average running torque, but the gasoline motor had to be started by hand and required a friction clutch which permitted it to be later placed in driving connection with the running gear of the car. Another deficiency of the gasoline motor was its inability to deliver excess power or to carry overloads for short periods. To overcome these objectionable features of the gasoline propulsion system it was proposed to combine electric and gasoline power, and a number of combination gasoline-electric models were built possessing the advantage of a self-starting power plant of considerable flexibility. The origin of electric self-starters for gasoline automobiles may be traced to those early combination vehicles, although the self-starting feature probably was rather secondary in the minds of the inventors, whose chief aim was greater flexibility of the power plant. The power plant generally consisted of a gasoline motor direct connected to a shunt-wound dynamo which was electrically connected to a storage battery. Below a certain critical speed of the engine the dynamo would act as a motor, drawing current from the storage battery and assisting the gasoline motor in propelling the car. Above this speed the dynamo would act as a generator and charge the battery. Several such self-starting combination power plants were built in the late nineties of the past century.

Electric lighting was used on electric vehicles from the beginning of the electric vehicle industry. Only carbon filament lamps were then available, which were rather inefficient, consuming about 3 watts per candle power. However, electrics needed comparatively little illumination, and as they carry a rather large store of electric energy this inefficiency did not stand in the way of electric lighting. On gasoline cars electric lights came to be used only after the advent of the metal filament (tungsten) lamp, which consumes only slightly more than 1 watt per candle power. This was about 1908. At first the lamps were supplied with current from a storage battery, which was recharged from outside sources when it was exhausted, but dynamos were soon developed which would automatically keep the battery charged. Among the first dynamo lighting systems to be brought out were the Apple and the Gray & Davis in 1909 and 1910. Once a regular electric generating plant for automobiles had been perfected, so that an ample supply of electric energy was constantly available on the car, there was nothing in the way of adding other electrically operated appliances, and one of the first to be added was the electric starter. The first combined electric lighting and starting system was marketed by the Dayton Electrical Laboratories Co. in 1911. Other systems soon followed.

Terminology—An electric starting and lighting system comprises two or three main elements and a number of subsidiary parts. There is first the electric generator which produces the electric energy consumed by the lamps and in starting the engine; then there is the storage battery in which the electric energy generated is stored until it is needed, and finally there is the electric motor which furnishes the mechanical power necessary to crank the engine. An electric generator and an electric motor are substantially one and the same machine—if we limit ourselves to continuous current machinery, which alone is used in automobile lighting and starting equipment. Stating the matter in another way, any electric generator will also serve as an electric motor, and vice versa. If the machine is driven from some source of mechanical power it generates electric current, and if electric current from an outside source is sent through it, it develops mechanical power. For that reason the same machine is sometimes used both as a generator to keep the battery charged and as a motor to start the engine. Such a double purpose machine is known as a dynamotor. The term dynamo is applied to both generator and motor. The reason that the same machine is not always used both as a charging generator and a starting motor is that if a separate machine is used for each of the two purposes it can be designed to perform its function more efficiently, as will be explained further on.

Principles of the Dynamo Generator—A dynamo generator operates on substantially the same principle as a magneto generator, the only essential difference between the two being that the field frame of the former is an electro-magnet, while that of the latter is a permanent magnet.

Dynamo generators, like magnetos, in the first place generate alternating current. The alternating current may be either used as such, or it may be transformed, or commutated, before it leaves the machine, into a direct or continuous current, this being accomplished by means of a device known as a commutator. In electric lighting and starting, since most of the current passes through a storage battery, only direct current can be used, and all dynamos of electric lighting and starting systems are of the direct current type and comprise a commutator.

The essential parts of a dynamo then are:

1. A field frame producing a magnetic field of force.
2. An armature carrying conductors in which an electromotive force is induced as the armature is revolved in the magnetic field.
3. A commutator which "commutates" the induced current, changing it from alternating into direct current.
4. A brush rigging and brushes for collecting the current from the revolving commutator.
5. Field windings through which an electric current is sent in order to establish a strong magnetic field.

Field Frames—Owing to the fact that lightness is an essential of all motor car parts, only materials of high magnetic permeability are chosen for the field frames of automobile dynamos, such as wrought iron, wrought steel or cast steel. Occasionally the field is built up of sheet metal laminations and this construction insures the lowest possible weight, as the sheet metal can be more thoroughly annealed and therefore has a higher magnetic permeability than solid blocks of metal. Magnetos used for automobile ignition have only two poles, which corresponds to a single magnetic circuit, but dynamo-generators often have four and sometimes even more poles. All or nearly all automobile dynamos are of the iron clad type, that is, there is an outer ring or frame from which poles extend inwardly. In Fig. 187 are shown some of the forms of field magnets most commonly employed. The one shown at A is a two-pole frame of substantially rectangular form. Field exciting coils are placed on both field poles. At B is shown a similar form, with one "salient" pole on which the field exciting coil or coils are placed and one "consequent" pole. C shows a four-pole frame of the round type and D a four-pole frame with two salient and two consequent poles. In all of these diagrams the paths of the magnetic lines are indicated by arrows. It may be pointed out that while in a magneto it is necessary to have the limbs of the magnets comparatively long, since the coercive force, which is the cause of the magnetic field of force, is proportional to the length of the magnets, in a dynamo the aim is to make the magnetic circuits as short as possible, as here the magnetomotive force originates in the exciting coils and an increase in the length of the magnetic circuit simply adds to the reluc-

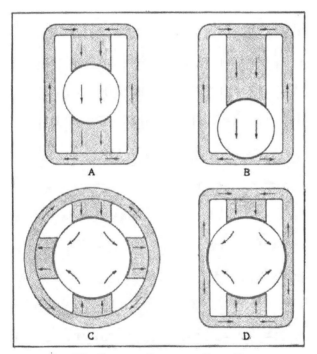

FIG. 187—TYPES OF GENERATOR FIELD FRAMES

tance. It is this consideration which leads to the adoption of multipolar field frames, as in many instances the length of the magnetic circuits can be shortened by using more than two poles.

In many cases the field poles are provided with pole shoes as shown in Fig. 188. The object sought is to increase the area of the air gap in order to reduce the magnetic reluctance of same, and yet keep down the cross sectional area of the

ELECTRIC GENERATORS

poles on which the exciting coils are wound. The magnetomotive force does not increase with the length of the field turns, whereas the amount of wire required and the energy consumption do increase with this length. These pole shoes also come in handy for supporting the field coils.

Principle of Commutator—The principle of the commutator is illustrated in Fig. 189 which shows a coil of wire which is supposed to rotate between poles of a two-pole magnet. Each end of the single turn is connected to a semicircular metal segment, and a brush (not shown) bears on each segment. The direction of the e.m.f. induced in the coil changes every time the sides of the coil pass out from under one pole and beneath the other, and as at the same time the brushes pass from one segment to the other, the polarity of the brushes always remains the same. If the brushes were connected by an outside wire, then a unidirectional, though pulsating, current would flow in the wire. The value of the current would constantly fluctuate between zero and maximum. By provid-

FIG. 188—FIELD POLE   FIG. 189—ARMATURE COIL

ing the armature with a considerable number of sections and the commutator with a corresponding number of bars, the fluctuations can be almost but not completely eliminated.

Commutator Construction—The commutator of an actual generator consists of a number of stamped copper bars or segments, equal to the number of sections in the armature, which are separated by sheets of insulating material, usually mica. The bars and interposed sheets of mica are firmly held together by a clamping device consisting of a metal sleeve with a head whose inner side is undercut at an angle, a washer similar in shape to the head of the sleeve and a nut screwing over the end of the sleeve. The sleeve is surrounded by a bushing of insulating material and washers of the same material are placed between the assembly of commutator bars and the two clamping heads. Each bar is then completely insulated from every other bar and from the clamping sleeve. A sectional view of such a commutator is shown in Fig. 190. At the rear end (toward the armature) the commutator bars are provided with lugs. These are slotted and into the slots the armature leads (ends of coils) are soldered. After being assembled the commutator is turned off smoothly all over and then sandpapered on the outer cylindrical surface, so that it presents a good bearing for the brushes. Sometimes the assemblage of copper bars and mica insulation, instead of being clamped between conical heads on a metal sleeve, is molded in synthetic insulating material such as bakelite or condensite.

Construction of Armature—The armature core is built up of soft sheet iron disks which are stamped with a hole at the center for threading them over the shaft and with slots all around the circumference in which the armature conductors are placed. Originally the conductors were wound on

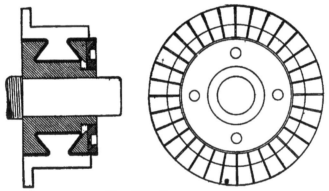

Fig. 190—Commutator

the outside of the core, but by placing them in slots the length of the air gap can be greatly shortened, whereby the energy required for exciting the field is reduced, and, besides, practically all driving strain is taken off the conductors and placed on the teeth of the core instead. The armature core must of necessity be made up of thin disks and these be slightly insulated from each other by varnishing them or placing sheets of thin paper between, because, if the core was one solid mass, eddy currents would be induced in it and result in great loss. The currents would flow in the core in the same direction as in the armature conductors, that is, toward the rear on one side and toward the front on the other. Since the core forms only a single conductor the e.m.f. induced in it is exceedingly low, and by also making the conductivity of the core low, by varnishing the disks, etc., the eddy current loss can be practically eliminated.

Fig. 191 shows an armature disk on which two forms of slots in common use have been drawn in.

The armature conductors are always put on in complete turns, and, since the electromotive forces induced in all parts of a section must be such as to act in the same direction along the length of the wire, the wire must be so wound on that when one side of a turn is, say, opposite a north pole the other side is opposite a south pole. The electromotive force induced in one side of the wire will then tend to send a current toward the rear of the armature and in the other toward the front of the armature, as indicated by arrows in

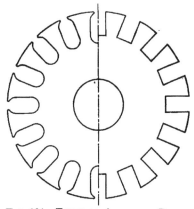

FIG. 191—TYPES OF ARMATURE DISKS

Fig. 192, and all induced e.m.fs. in the same armature section tend to cause a current to flow around the circuit in the same direction. From the above it follows that in a two-pole dynamo the two sides of an armature convolution must be located in diametrically opposite slots; in a four-pole dynamo they must be located in slots 90 degrees apart; in a six-pole dynamo in slots 60 degrees apart, and so on. Each slot generally contains halves of two armature sections, so that there are as many armature slots as there are armature coils or sections, but sometimes in small machines more than one section are placed in one slot.

In each section of the armature winding an electromotive force will be induced which, as the armature revolves, varies substantially according to the law of sines. The induced electromotive force is greatest when the conductors are in the middle of the pole shoes, as lines of force are then cut at the most rapid rate, and it is zero when the conductors are midway between pole shoes. In this latter position the direction of the current through the armature sections reverses.

After all of the sections have been wound on and the armature has been suitably treated with an insulating compound and baked, the end of one section is connected to the beginning of the next or adjacent section. This connection is usually made by soldering the end lead of one section and the beginning lead of the next to the same commutator bar. This is done all around the armature. The arrangement of circuits, in a two-pole armature, is therefore as diagrammaticaly indicated in Fig. 193. It will be understood, of course, that each

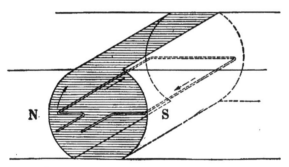

FIG. 192—INDUCTION IN ARMATURE COIL

coil or section is wound completely around the core, its turns passing from front to back through a slot on one side, and from back to front through a slot on the opposite side. In the diagram the coils are merely indicated by zigzag lines on that side of the core where the leads or ends are brought out. In the coils midway between pole pieces no electromotive force is induced. The electromotive forces in the sections to one side of these neutral sections are all in the same direction and those in the sections on the opposite of the neutral sections are all in the opposite direction. The whole armature winding therefore forms two parallel circuits, as indicated by arrows in Fig. 193.

Then, if brushes are placed on the commutator at B and B', and connected through an outside circuit, current will flow through the armature winding as indicated by the arrows and return through the outside circuit. As the armature rotates, sections of the winding are continually transferred, as their commutator bars pass under the brushes, from one group to the other, but at all times the electromotive forces of all sections in either group are equi-directional and therefore add together. Each armature section, of course, carries only one-half the current flowing in the outside circuit. Each section, as the commutator bars to which its leads are connected pass under the brush, is momentarily short circuited. It has already been stated that in one particular position no electromotive force is induced in the coils. If a coil is short circuited by the brush

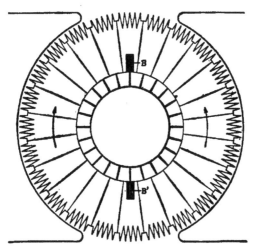

FIG. 193—GROUPING OF ARMATURE COILS

while an e.m.f. is induced in it a short circuit current will, of course, flow in the section. The conditions of commutation are best if this short circuit current at the moment the short circuit is broken is exactly equal to the current which will flow in the section when it is included again in one of the regular armature circuits, for then there will be no change in the value of the current flowing in the section as its short circuit is broken, and hence there will be no sparking at the brushes.

Hand and Form Wound Armatures—Armatures may be either hand wound or form wound. The first method is generally employed where small wires and many turns per section are used, the latter in the alternate case. A hand wound armature is illustrated in Fig. 194, and hardly anything need be said to explain the method. The winder begins at any particular pair of slots and winds on one section, then goes to the pair adjacent to these slots, and so on all around the armature core. One disadvantage of this method is that the first half of the sections wound on lie in the bottoms of the slots and the last half in the tops, hence not all sections are alike, as they preferably should be. This objection is overcome in the form wound armature as illustrated in Fig. 195. View A gives an idea of the appearance of the coils at the rear end and view B indicates how one-half of the end portion of the coil is made to extend out farther from the core to clear the other half. One side of a form wound section always lies in the lower part of the slot and the other side in the upper part and all sections are absolutely alike. The sections of a form wound armature are wound on a wooden form, taped, im-

FIG. 194—HAND-WOUND ARMATURE

pregnated with insulating varnish, and baked before being applied to the core. Hand wound armatures are impregnated and baked after they are completely wound and before the commutator is put on, or at least before the leads are soldered to it.

As the armature in service revolves at high speed, the centrifugal force acting on the armature coils tends to make them fly out of their slots. To prevent this, bands of piano wire are wound over the completed armature in depressions in the core provided for the purpose. In form wound armatures such bands are also wound over the ends of the armature coils that project beyond the core. First a band of mica strips is applied over the coils and then piano wire is wound on. Several sheet copper clips are placed under the band and their ends are bent over and soldered to secure the ends of the binding wire. In armatures whose slots are partly closed the "banding" may be dispensed with. Strips of fiber are then placed in the top of the slot, which close the opening and are held in position by the baked insulating compound.

ELECTRIC GENERATORS

Armature Winding Diagrams—Fig. 196 shows a winding diagram for a two-pole hand wound dynamo. The circles near the circumference of the armature represent the armature conductors. A cross (tail end of an arrow) in one of these circles is intended to mean that the current in the conductor flows away from the observer or from the front to the back of the armature, whereas a central point is intended to mean that the current in the conductor flows toward the observer or from back to front of the armature. Only the connections at the forward or commutator end of the armature are drawn in, the connections at the rear end being always from one slot to the diametrically opposite one. The beginning lead of each armature section is brought straight out to the commutator bar directly opposite it, while the end lead, which comes out exactly half way around the armature from the beginning lead is carried around the armature at the commutator end

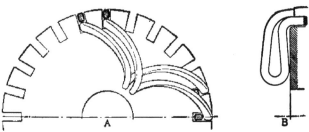

FIG. 195—SKETCHES ILLUSTRATING FORM WINDING

to the bar adjacent to the one to which the beginning lead is soldered. That it is a hand wound armature is shown by the fact that half of the beginning leads are at the bottom and half at the top of the slots. In a form wound armature all beginning leads would be in the same position in respect to the slot.

Fig. 197 is a diagram of a four-pole armature winding requiring the use of four brushes. Here the connections at the rear of the armature are also shown, and it will be seen that the two sides of each coil are a quadrant apart. This diagram applies to a form wound armature, as each coil comprises one bottom and one top section. The two positive and the two negative commutator brushes are connected together respectively. By tracing out the armature circuits it will be found that the end lead of each coil or armature section connects to the commutator bar adjacent to the one to which the beginning lead of the same coil connects, the same as in the two-pole winding shown in Fig. 196.

In some cases, especially where the machine will be located in a cramped space, it is desirable to use only two brushes on a four-pole machine, and it is then necessary to use a special method of connecting the armature windings, known as a series grouping, which is illustrated in Fig. 198. In this case the end lead of one section, instead of being connected to the commutator bar adjacent to the one to which the begin-

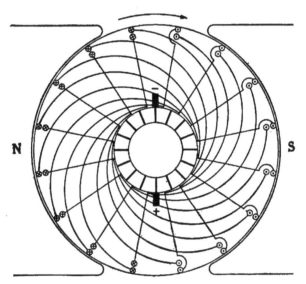

FIG. 196—DIAGRAM OF TWO-POLE ARMATURE WINDING

ning lead of the same section is connected, is connected to the bar directly opposite to the latter. With such a winding for a four-pole machine the number of sections must be odd.

Self Excitation—The field frame of a dynamo being an electromagnet, it loses practically all of its magnetism when the machine is stopped. When it is started again the small amount of remanent magnetism of the field causes a small electromotive force to be induced in the armature windings. This enables the armature to furnish current for the field coils, and the fields will thus be strengthened. The e.m.f.

induced in the armature will then increase, and so in turn will the strength of the field, this gradual growth of armature e.m.f. and field strength continuing until a condition of stability is attained. The phenomenon just described is known as "picking up." Under certain conditions, as if the commutator is dirty, so that the brushes are not making good electrical contact, the machine may fail to pick up, but when everything is right it will always do so.

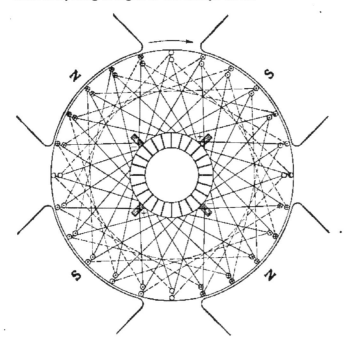

Fig. 197—Diagram of Four-Pole Armature Winding, Parallel Grouping

Methods of Field Connection—The field magnets of a dynamo, of course, may be excited from an outside source of current, but this plan is not employed in actual work. The energy required for field excitation amounts to only a few per cent. of the electrical energy generated by the dynamo, and part of the latter energy is therefore used for the fields. There are several different methods of connecting the field coils to the armature, and the characteristics of the dynamo vary greatly with these. In the first place, we may wind the fields with a relatively small number of turns of heavy wire and send the whole of the current derived from the armature through the field winding before it is sent to the consuming devices (see Fig. 199). The field coils are then connected in series with the armature, and such a machine is known as a series dynamo. This type, however, is practically never used as a generator, for the reason that its voltage varies greatly

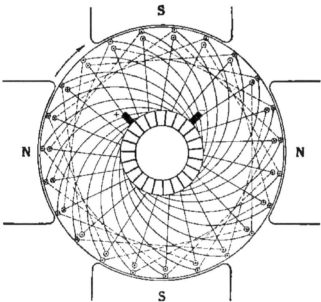

FIG. 198—DIAGRAM OF FOUR-POLE ARMATURE WINDING, SERIES GROUPING

with the load, whereas in practically all applications of dynamos a substantially constant voltage is required. This may be explained as follows: Suppose that the dynamo carries only a light load; in other words, that the resistance of the circuit is high. Then only a small current will be delivered by the dynamo, and as it is this current which flows through the field coils, the excitation will be low, there will be a relatively small magnetic flux, and consequently a low induced e.m.f. If now the load on the dynamo is increased by lowering the resistance of the outside circuit, then the main current will increase, and with it the field excitation and the induced e.m.f.

By the second method of excitation, known as shunt excitation, the field circuit is connected in shunt across the armature, as shown in Fig. 200. In that case the current flowing through the field coils is always proportional to the e.m.f. of the armature, and is almost wholly independent of the load carried by the dynamo. The characteristic curves of both shunt and series dynamos are shown in Fig. 201, from which it will be seen that the e.m.f. of the shunt dynamo is nearly constant throughout the whole range of the load, there being a slight drop in e.m.f. with an increase in load. This is due

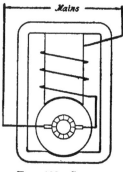

FIG. 199—SERIES
GENERATOR

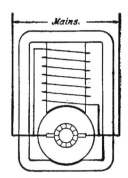

FIG. 200—SHUNT
GENERATOR

to the fact that there is some loss in voltage in the armature windings when the latter carry current, and to the further fact that the armature current distorts the magnetic field and reduces its useful component.

When an absolutely constant voltage is required, as in house lighting, the dynamo is provided with both a shunt and a series field winding, or what is known as a compound winding, as shown by Fig. 202. It has been pointed out that with a shunt winding alone the voltage tends to drop slightly as the load on the dynamo increases. Now, the few series field turns on a compound wound dynamo, of course, increase in strength as the load rises, and compensate for the drop mentioned, thus keeping the voltage constant. Diagrammatic representations of series, shunt and compound dynamos are shown in Fig. 203.

All the preceding considerations apply to conditions of constant speed, under which conditions nearly all stationary motors are operated. But the automobile engine operates at variable speed, and in the case of a dynamo driven from it the problem of control must be solved along different lines. With an increase in speed both the e.m.f. and the current delivered by a dynamo tend to increase. Now, in all automobile electric systems a storage battery is constantly connected to the dynamo as long as the e.m.f. of the latter is sufficient for charging purposes, and to prevent charging at an unduly high rate when the engine runs fast, some means must be

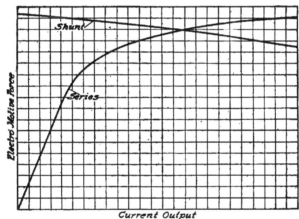

FIG. 201—CHARACTERISTIC CURVES OF SHUNT AND SERIES GENERATORS

adopted to prevent this increase in output with an increase in speed, or at least to minimize it. One way of accomplishing this is to provide the dynamo field with a shunt winding and a reverse or opposing series winding, the two together forming a differential winding. Then, as the e.m.f. and current delivered increase as the result of an increase in speed, the current through the reverse series winding also increases, and this reduces the strength of the magnetic field. This weakening of the magnetic field compensates largely for the increase in speed, and there will therefore be only a slight gain in output with an increase in speed.

The characteristics of the four different types of dynamo generators therefore may be briefly recapitulated as follows:

ELECTRIC GENERATORS

Series dynamo—at constant speed e.m.f. grows rapidly with output.

Shunt dynamo—at constant speed e.m.f. is nearly constant, dropping only slightly with an increase in output.

Compound dynamo — at constant speed gives an almost absolutely constant e.m.f. for any load within its capacity.

Differential dynamo—tends to keep output constant despite variations of speed.

Brushes and Brush Rigging—The brushes which bear on the commutator and take off the current are generally made of carbon or graphite—in rare

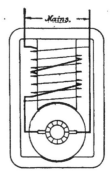

FIG. 202—COMPOUND GENERATOR

instances of copper. They are pressed against the commutator by a spring. Almost any form of spring may be used, from a coiled compression spring to a flat leaf spring. Since the spring presses directly on the current-carrying brush, its support or anchorage must be insulated from the dynamo frame. In fact, the spring forms a conducting connection between the movable brush and the stationary current-carrying part, and may serve to carry the current from the brush to the dynamo terminal. However, where appreciable currents are involved, it is safer to provide a flexible cable connection, known as a pig tail, between the brush and the holder, as if the spring has to carry most of the current it is liable to become hot and lose its temper. A well-designed brush rigging is illustrated in Fig. 204. The brush holders A, which contain the carbon brushes B, are secured to but insulated from a rocker ring C, which fits over the bearing hub of the dynamo. A shear type wire spring D containing a number of turns is secured to a lug on the brushholder bracket, and its free end extends into slots in the holder and presses on the carbon. The end of the spring is looped so it can be easily grasped between the

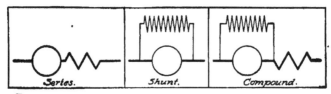

FIG. 203—DIAGRAMS OF THE DIFFERENT TYPES OF GENERATORS

fingers when it is desired to renew the carbon. Flexible cables E connected to the carbons are provided with cable terminals at their ends, and connect to terminal posts in the housing of the dynamo. In this particular design there are two brushes side by side (this cannot be seen in the drawing). This feature tends to reduce sparking, because if the contact of one brush with the commutator should be momentarily impaired by a slight unevenness of the bearing surface, most likely the other brush would make good contact at the time.

In the brush rigging shown in Fig. 204 the brush follows

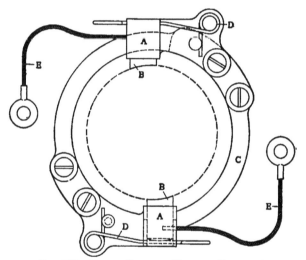

FIG. 204—BRUSH RIGGING (SLIDING BRUSHES)

the outline of the commutator, and is fed toward same as it wears, by sliding in the brush holder. There is another type of brush rigging in which the brush holder has a pivot support around which it swings, the brush being securely clamped in the holder. The latter type is exemplified by the Delco construction shown in Fig. 205. Here the free end of the holder is drawn toward the commutator by a coiled spring surrounding the pivot stud, one end of the spring being anchored on a stationary pin and the other end hooked over a pin on the brush holder. A flexible cable is secured to the brush holder, so the current does not have to pass through the bearing surface of the pivot support.

ELECTRIC GENERATORS 253

Commutation—One of the prime essentials of a successful vehicle dynamo is sparkless commutation. As long as the dynamo carries no load the position of the brushes ensuring commutation under the most favorable circumstances is such that the coil short-circuited by the brush spanning two commutator bars lies substantially midway between pole pieces, this being the position of minimum induction. If we shift the brushes from the position in either direction the e.m.f. between brushes will decrease until when the brushes have been shifted through an angle equal to one-half the angle

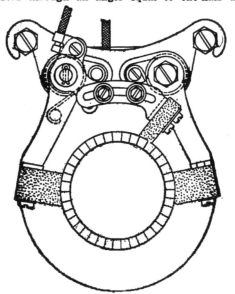

FIG. 205—BRUSH RIGGING (ROCKING BRUSH HOLDERS)

between the centers of adjacent poles there will be no e.m.f. between them. At the same time there will be severe sparking at the brushes.

When the dynamo is running under load and the armature consequently carries considerable current, a cross magnetic field is set up by the armature current which tends to shift the position of most favorable commutation—forward in the direction of armature rotation in the case of a generator, and backward in the case of a motor. The explanation of this phenomenon is that the magnetic field due to the armature current is at right angles to the regular magnetic field, and the two together form a resultant which is angularly displaced with relation to the magnetic field of the field frame. This, of course, shifts the position of "no induction," and hence the line of commutation. Theoretically, therefore, the brushes would have to be shifted slightly around the commutator as the load increased and decreased. In practice this obviously cannot be done, and the aim then must be to reduce the effect of "armature cross magnetization" to a minimum. One way of accomplishing this is to make the magnetic field relatively strong, and put comparatively little copper on the armature, which, however, leads to a rather heavy and expensive machine for a given output. Another method sometimes applied consists in forming the pole pieces with a rather deep and wide lengthwise central slot. This, as indicated in Fig. 206, introduces a gap in the path of the cross-magnetic lines, and therefore reduces their number.

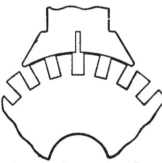

FIG. 206—SLOTTED POLE SHOE

Prevention of Humming at High Speeds—As the teeth of a toothed armature pass out from under the pole face there is a sudden change in the magnetic flux through them, and this, at high speed, gives rise to a somewhat unpleasant humming noise. To obviate this noise the design must be so arranged that not the whole tooth passes out from under the pole face all at once. This can be accomplished either by "twisting" the armature core so that the armature slots, instead of running parallel with the armature axis, form parts of helices on the armature surface; by similarly twisting the pole faces or by making the edges of adjacent pole shoes approach each other more closely at the middle of their length than at their ends.

Capacities of Automobile Generators—One of the most difficult problems in connection with the design of a generator for an automobile electric system is to determine the best output capacity. Not only does the amount of electrical energy consumed vary somewhat with the size of the car and with its equipment, but it varies greatly even with the same make of car at different seasons of the year and for different conditions of operation. The generator furnishes current continuously while the engine is running above a certain minimum speed. On extended drives in the open country, as in touring, it will be sending current into the battery continuously for long periods. Current is taken out of the battery for different purposes, such as lighting, starting, operating the horn, etc. In touring, very little is generally required for engine starting, while the amount required for lighting depends upon the amount of night driving done. Under average touring conditions there is generally much charging and little discharging, for the average tourist, after having driven more or less all day does not care much for night driving in strange country. On the other hand, when a car is used principally in the city, by a man who is occupied indoors during the day, there is apt to be a great demand for current for lighting and little opportunity for recharging. In such cases the battery is generally in a very low state of charge, and deteriorates quickly.

Both overcharging and undercharging are objectionable, though the former is not nearly so injurious as the latter. It is therefore well to select a generator capacity which will insure a certain amount of overcharging under average conditions. The largest generators used on cars have a maximum capacity, or are regulated for a maximum capacity, of about 160 watts, while most generators have an output of from 100 to 120 watts.

Care of Dynamos—The majority of lighting and starting dynamos are equipped with ball bearings, but a few have plain bearings on the armature shaft. The armature bearings and the commutator, with its brushes, are the only wearing parts of the dynamo, and therefore the only parts requiring periodic attention. Bearings must be oiled at intervals, and in regard to the length of these intervals it is best to follow the maker's instructions. One of the chief points to be observed in the case of automobile dynamos is to avoid overlubrication. This is due to the fact that oil accumulations are an enemy of good insulation. Oil always collects dust, and the dust inside a dynamo of the enclosed type consists mainly of particles of conducting material worn off the commutator and brushes. If much excess oil gets into the machine, conducting layers formed of oil and metal or carbon dust are apt to form, and short-circuit the machine. On the other hand, oil also attacks rubber insulation with which the flexible cables between the brushes and the terminal posts are usually covered. If oil collects on the commutator it interferes with good brush contact and induces

256   ELECTRIC GENERATORS

sparking. Dirt on the commutator is indicated by tiny sparks encircling the commutator. It must be removed by wiping the commutator with a dry cloth while the machine is running. Another trouble likely to result from overlubrication is that the oil gums the carbon brushes tight in the brush holders, which impairs the contact between the brush and the commutator and results in sparking. When this occurs, the remedy is to remove the brushes from the holders and thoroughly clean the contacting surfaces of both parts.

Care and Adjustment of Brushes—For sparkless commutation accurate positioning of the brushes is most essential. In a two-pole machine the two brushes must be located exactly diametrically opposite each other; in a four-pole machine, exactly at quarters, etc. Most machines are provided with means for adjusting the brushes, and if their adjustment is believed to have been disturbed, as in the case of repair work, the position of most nearly sparkless commutation should be found by trial. Another essential to good commutation is the right amount of spring pressure on the

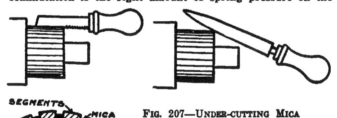

Fig. 207—Under-cutting Mica

brushes. If the springs are too light the brushes are apt to shatter, especially where the whole machine is subject to considerable vibration, as on a car. On the other hand, if the pressure is too great there will be considerable loss due to brush friction, with the resultant heating, and both brushes and commutator will wear away quickly. Long experience has proved that a spring pressure of about 2 pounds per square inch of brush-contact surface is good practice. The experienced electrician can tell whether the pressure is substantially right by simply lifting up on the springs.

Sometimes one or more of the mica plates which separate adjacent commutator bars will come loose and rise slightly above the surface of the commutator. The protruding mica can generally be removed by means of a sharp knife. After long use the commutator generally becomes rough, because the mica is harder than the copper, and wears away more slowly. The armature must then be removed from the dynamo and placed in a lathe, and the commutator turned down, taking off just sufficient to remove all unevenness, and then smoothing it with fine sandpaper. In its normal condition after a period of use the commutator presents a rich, glossy brown color. If the color is of a lighter shade a little lubricant will improve conditions. There are various "commutator compounds" on the market, but in the absence of these a stick of paraffin wax will do. It is applied to the surface of the commutator while the armature is in motion.

Undercutting Mica—Where brushes of relatively soft material are used there is a tendency for the copper bars to wear down faster than the mica, causing the mica to project from the commutator, and resulting in arcing at the brushes. To prevent this the mica is undercut so that the brushes never touch it. The commutator comes with the mica undercut, from the factory, but after the copper has worn down somewhat the mica must be undercut or slotted again. The following instructions for doing this are from the literature of the Dayton Engineering Laboratories Co.

The armature is first removed from the machine and placed in a lathe, where the commutator is trued up so as to be perfectly concentric. This work should be very carefully done, and as fine a cut as possible taken, to avoid waste of copper. When the commutator has been trued up, cut out the mica between the copper bars with a hacksaw blade, the sides of its teeth having been ground off until it will cut a slot slightly wider than the mica insulation. In this way a rectangular slot free from mica, approximately 1/32 inch in depth, is obtained between two adjacent commutator bars. After undercutting the mica the edges of these slots should be beveled very slightly with a three-cornered file, in order to prevent any burs remaining, which would cause excessive brush wear. This method of undercutting the mica puts the commutator in good condition. Reference to Fig. 207 will show how the slots should look when finished.

Sanding-In Brushes—When new brushes are fitted they must be bedded to the commutator. In some cases the brushes are arranged to stand radially to the commutator, but generally they stand at a slight angle, so that the friction between the brush and commutator tends to increase the pressure between them. The brush is first ground or filed down on its bearing surface to the approximate angle of contact. Then it is placed in position in the brush holder, a sheet of fine sandpaper is placed on the commutator with the back or paper side toward the latter and the "business

258   ELECTRIC GENERATORS

side" toward the brush, and while one man presses down on the brush another draws the sandpaper back and forth over the surface of the commutator. In this way the brush is soon worn down to a cylindrical surface conforming to the wearing surface of the commutator.

The Dayton Electrical Laboratories Co. gives the following directions for "sanding in" brushes: Use a fine piece of No. 00 sandpaper, cut in strips slightly wider than the brush. Never use emery cloth. The sandpaper is wrapped around the commutator so as to make contact with at least half of the circumference, as illustrated in Fig. 208, with the sand side away from the commutator and in contact with the end of the brush. Now, by drawing this strip of sandpaper back and forth it is possible to fit the brush very accurately to the commutator.

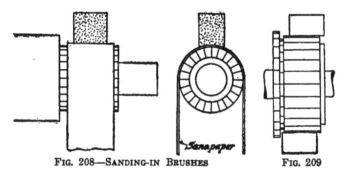

FIG. 208—SANDING-IN BRUSHES   FIG. 209

Small unevennesses which form on the commutator in the course of time can be removed by means of fine sandpaper which is applied to the commutator while the machine is in operation. To prevent shoulders being worn on the commutator at the sides of the brushes the latter may be staggered so that one brush slightly overlaps the outer edge and the other the inner edge of the commutator, a groove being turned in same adjacent to the lugs to which the wires are soldered, so as to form a sharp ledge at the inner end of the bearing surface. (See Fig. 209.) This practice is especially to be recommended where copper gauze or leaf brushes are used, as they wear the commutator more rapidly than carbon brushes.

CHAPTER XXIII

Generator Control

All automobile electric generators are primarily shunt wound machines—i.e., the main field winding is a shunt winding. Such a machine has a speed-voltage characteristic of the general form shown in Fig. 210, the resistance of the outside circuit being supposed to be constant and sufficiently large so that at normal speed the current will not exceed the rated output capacity. As has been explained before, the voltage induced in a generator armature is in direct proportion to the speed of revolution of the armature, the number of magnetic lines of force passing through it and the number of conductors on the armature that are connected in series. As long as one of these factors, the speed, is very small, the voltage cannot be great, the more so because one of the other two factors, the number of lines of force, also increases with the speed. This is so because the voltage increases directly with the speed, the field exciting current is directly proportional to the voltage and the number of lines of force increases with the exciting current. The voltage therefore increases rapidly with the speed, as the diagram shows.

Need of Battery Cutout—As long as the voltage of the generator is below that of the battery, the former must not be connected to the latter, because if it were, the battery would discharge through the generator armature and field coils. Therefore, most automobile electric systems include an automatic switch or battery cutout which connects the generator to the battery as soon as the generator voltage attains a value sufficiently high for charging, and disconnects it from the battery when the voltage falls below this value. Generators and their drives are generally so designed that this voltage is attained at a comparatively low car speed, from 7 to 10 miles per hour on the direct drive. As the speed continues to increase, the generator voltage continues to rise, and since it requires only a small increase in charging voltage to produce a relatively large increase in charging current, with a positively driven generator of the shunt wound

260 GENERATOR CONTROL

type a point would soon be reached where the battery would draw an excessive current—to its own detriment as well as that of the generator.

Constant and Variable Speed Generators—In stationary electric plants the generators are always driven at constant speed, or as nearly constant speed as possible, and the problem of regulation is then a comparatively simple one. For electric lighting service, for instance, a substantially constant line voltage is required, irrespective of the number of lamps in circuit, and this is obtained by providing the generator field with a main shunt winding and a small supple-

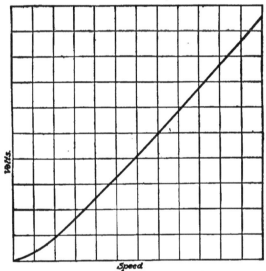

FIG. 210—SPEED-VOLTAGE CHARACTERISTIC OF SHUNT GENERATOR

mentary series winding, which latter compensates for the drop in voltage with increasing load observed in a plain shunt-wound dynamo. Not much had been done in the way of developing variable speed generators for battery charging previous to the advent of dynamo lighting systems for automobiles, though several systems of train lighting by means of current from generators driven from the car axles had been developed, and some attempts had been made to use windmills or aerial motors for generating current for storage battery charging.

Battery Floating on the Line—In Fig. 211 is shown a simple diagram of connections which is typical of all automobile

GENERATOR CONTROL 261

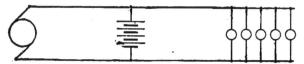

FIG. 211—BATTERY FLOATING ON THE LINE

electric systems. From the terminals of the generator, connection is made through the automatic switch to the battery, and from the battery connection is made to the consuming devices—such as lamps, starter, etc., through hand or foot-operated switches. In such installations the battery is said to "float on the line." Current can flow from the generator direct to the consuming devices, and the battery will either receive or deliver current, depending upon the amounts generated and consumed, respectively. If more is generated than is consumed the difference goes into the battery, and if more is consumed than is generated, the difference comes from the battery. The voltage applied to the consuming devices is governed by the battery. In other words, it is substantially that at the terminals of the battery, and as long as no excessive current flows into or out of the battery the voltage applied to the consuming devices will range within permissible limits. The problem of generator regulation therefore consists chiefly in assuring that the current passing into the battery never exceeds a certain value.

Voltage and Current Control—There are two distinct principles of regulation, viz., regulation for constant voltage and regulation for constant current. The respective merits of these principles depend chiefly upon their effect on the charging of the battery. If constant current regulation is used, then, as long as no consuming devices are connected

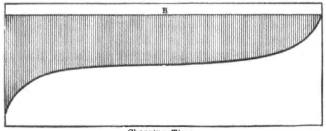

FIG. 212—TAPERING CHARGE

to the battery, the latter receives a charging current of constant value, whether it is already fully charged or whether it is nearly discharged. With constant-voltage regulation, on the other hand, the charging rate varies with the state of charge of the battery, being greatest when there is very little charge in the battery and decreasing as the amount of charge in the battery increases. This may be explained as follows: The electromotive force of the battery rises with an increase in the amount of charge in the battery, as shown by curve A, Fig. 212. Straight line B represents the constant voltage of the generator and the current flowing into the battery, for any state of charge of the latter is proportional to the distance between curve A and line B, along the vertical line corresponding to this state of charge. Therefore, as the charge in the battery increases, the current flowing into the battery will decrease as the height of the shaded area in Fig. 212. This gives what is known as a "tapering charge" which is generally recommended by battery manufacturers. One of its advantages is that if by a heavy drain, as by extended night driving, the battery has been brought to a low state of charge, the generator will send a comparatively heavy current into the battery, thus filling it up rapidly, while if the battery holds a complete or nearly complete charge only a very small current is sent into the battery. If the battery is already fully charged all of the current sent into it is wasted, and it is, of course, desirable that this waste be as small as possible. There is no doubt that the tapering charge is the most efficient for any state of charge. On the other hand, there is one important thing to be said in favor of constant-current regulation. Suppose that the battery has been neglected and in consequence is badly sulphated, so that its internal resistance is comparatively high. Then, with a constant voltage system it will receive a very low charging current, which will tend to keep it in a state of low charge and therefore aggravate the sulphation instead of curing it. On the other hand, with constant-current regulation the charging current sent into the battery will be the same, whether the battery is sulphated or not.

Methods of Voltage Control—Since the voltage depends upon three factors, viz.,

(1) Speed of revolution;
(2) Number of lines of force;
(3) Number of effective armature conductors,

we may obtain a constant voltage

(1) By keeping the speed of revolution of the armature constant;

GENERATOR CONTROL

(2) By decreasing the number of lines of force to compensate for increase in the speed of the armature;

(3) By decreasing the number of effective armature conductors to compensate for an increase in the speed of the armature.

Since the engine speed varies greatly with the car speed and with changes of gear, if the generator speed is to remain constant, we have to employ some form of clutch which will slip when the engine runs beyond a certain speed. Such "governor clutches" have been employed by several well-known makers of electric equipment both here and abroad,

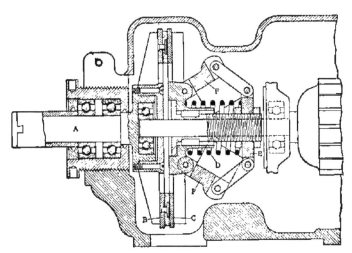

FIG. 213—GRAY & DAVIS GOVERNOR CLUTCH

including Gray & Davis, Hartford, Peto & Radford. A good example of such a clutch is the original Gray & Davis, illustrated in Fig. 213, which was used for several years.

Governor Clutch—Referring to the illustration, A is a driving shaft which is driven positively from the engine crankshaft and whose speed therefore increases and decreases with that of the latter. To this shaft is secured an aluminum friction disk B, which is faced with asbestos fabric. Facing this friction disk is another, cast iron one, C, which is constrained to turn with the armature shaft, but capable of moving lengthwise thereto. The two disks are pressed together by a coiled spring D which rests against an adjustable collar

E, on the armature shaft. Hinged to disk C are two governor weights FF, and these are also connected by links to the collar E. At low speeds of the engine the spring presses the two disks together with sufficient force to insure the armature being carried along with the driving shaft at the same speed as the latter, but above 1200 r.p.m. the centrifugal force on the governor weights causes the latter to move out from the axis of rotation against the pressure of the spring, thereby momentarily drawing the driven disk away from the driving disk and causing slippage between the two. As soon as the armature slows down the centrifugal force on the governor weights diminishes and the spring presses the disks into contact again. The pressure of the spring is so adjusted by means of collar E that the armature will be kept running at 1200 r.p.m.

Under ordinary conditions of driving the clutch will naturally be slipping more or less all the time, consequently will generate heat and produce wear. In order to reduce the wear to a minimum the friction surface is made comparatively large, so that the pressure on it per unit of area is small (about one pound per square inch). To keep down the heating, both discs are provided with fan blades, and there are ventilating openings provided in the disks and in the housing of the generator.

One objection to the constant speed or governor pulley system is that it is not very efficient. Suppose, for instance, that the gearing is such that at a car speed of 12 miles per hour the dynamo armature runs at 1200 r.p.m. and the clutch, therefore, just begins to slip. Then at 24 miles per hour the clutch will be slipping 50 per cent—i.e., its driven member will be running just half as fast as its driving member, and just as much energy as is transmitted will be lost in friction. In other words, the efficiency of transmission is only 50 per cent. Inasmuch as the power transmitted is only small (less than ¼ hp.), this is not a serious matter, yet it is one point in respect of which the governor clutch method of control compares unfavorably with other methods.

The electrical conditions with this system of control are ideal. By using a properly compounded dynamo the voltage is maintained almost absolutely constant and the battery receives a tapering charge.

Control by Changing Field Strength—The second method of control consists in reducing the number of magnetic lines through the armature as the speed increases. This can be done in several ways as follows:

(1) Resistance may be introduced in the field circuit by

GENERATOR CONTROL

some automatic device actuated by current from the armature proportional to the generator voltage.

(2) The field strength may be changed by including in it a resistance controlled by a magnetic vibrator which will pass less current when it vibrates more rapidly:

(3) The field may be provided with a reverse series coil;

(4) The armature may be partly drawn out of the "armature tunnel" at high speed by means of a centrifugal governor;

(5) A magnetic shunt may be provided across the pole pieces which is drawn closer as the field strength increases.

Variable Resistance in Shunt Circuit—Examples of the first method of regulation are to be found in the control of the Bosch system and of a former U. S. L. system. The

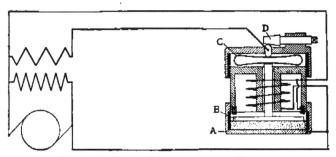

FIG. 214—BOSCH VOLTAGE CONTROL SYSTEM

Bosch control system is illustrated in Fig. 214. It will be seen that there is a small auxiliary shunt field winding, which is permanently connected to the armature terminals, and another, main shunt field winding, which includes a variable resistance, in circuit. This variable resistance consists of a mixture of ground carbon and mica, the mica being used to impart a certain "springiness" to the mass. It is placed in a corrugated metal cup A and a perforated steel plate B bears on top of it, the pressure on the plate being due to the spring C. Plate B forms the armature for an electromagnet whose coil is connected in shunt across the generator terminals.

Carbon is an electric conductor, but mica is an insulator and as the carbon particles when not subjected to any pressure are in loose contact, the mass of the mixture offers considerable resistance to the passage of the current. If pressure is applied to the mass the carbon particles come into more intimate contact and the resistance is consequently reduced.

It will be understood that as the electromotive force developed by the generator increases by reason of an increase in speed, the electromagnet becomes stronger and attracts armature B against the pressure of spring C. This lessens the pressure on the resistance material and increases its resistance, because with reduced pressure there is less intimate contact between the carbon particles. But as the resistance of the carbon increases the current through it and the main field winding decreases, thus decreasing the strength of the magnetic field, which tends to keep the induced voltage constant. By means of cone D the pressure of spring C can be varied and thus the rate of charging be changed to suit different conditions.

The principle involved in the old U. S. L. system was substantially the same, only instead of a ground carbon and mica mixture there was used a pile of carbon disks which were pressed together by a spring and the pressure was eased by an electromagnet. Only a single shunt field winding was used and the carbon disk regulator was connected in circuit with this, but in addition there was a relatively weak reverse series winding.

Solenoid-Controlled Field Rheostat—Instead of varying the resistance by exerting more or less pressure upon a "springy" conducting mass, a regular multi-section rheostat may be used which is actuated by a solenoid connected either in series with or in shunt to the main line. This plan is followed in the Adlake system, of which a diagram is shown herewith. The generator is shunt wound and the rheostat is connected in series with the shunt field winding. In the illustration the generator, battery and lamps are shown diagrammatically, while of the regulator itself a regular mechanical elevation is shown, parts being sectioned to disclose the interior construction. Referring to the cut, the solenoid coil is shown at G, and within this coil there is an iron core K, which connects by means of a cable L over a pulley M with the piston N of a dash-pot O. The magnetic suction of solenoid G on core K is counter-balanced by the weight of piston N and a certain number of shot contained therein. When the charging current increases by reason of an increase of the speed of the gasoline motor, solenoid G exerts a stronger pull on core K. This causes the pulley M to turn on its axis, and since rheostat arm F is rigidly fastened to the pulley, more sections of the rheostat are thereby cut into the field circuit. This reduces the current flowing through the field coil and

GENERATOR CONTROL

hence the field strength, and thereby makes up for the increase in speed. By increasing or decreasing the number of shot in the piston end of the dash-pot, the output for which the regulator is set can be varied at will.

Generator Output Increased for Lamp Load—The Adlake regulator embodies a feature also found in several other systems—viz., that of. automatically increasing the current output of the generator when the headlights are turned on. To this end the winding of the solenoid is made in two sections. When the headlights are off the current flows through

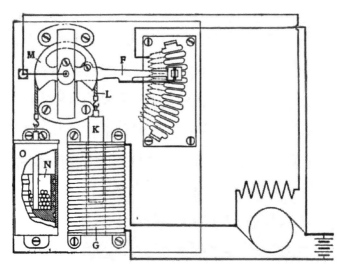

FIG. 215—ADLAKE CURRENT REGULATOR

the whole of the solenoid winding, but by turning on the headlights, part of the solenoid winding is cut out of circuit, and since the solenoid has then fewer effective turns, a stronger current must flow through it to move the rheostat arm to a certain position.

Mercury Tube Regulator—Still another embodiment of the system of controlling the voltage of the generator by varying an outside resistance in the shunt field circuit is found in the voltage regulator employed with Delco systems for several years. A sectional view of this regulator is shown in Fig. 216.

It comprises a solenoid A surrounding the upper half of a mercury tube B. Within this mercury tube is a plunger C consisting of an iron tube with a coil of resistance wire wrapped around its lower portion on top of a layer of insulating material. One end of this resistance wire is connected to the lower end of the tube and the other end to a pin D in the centre of the plunger. The lower portion of the mercury tube is divided by an insulating tube into two concentric wells, the plunger tube being partly immersed in the outer well and the pin in the inner well. The space in the mercury tube above the mercury is filled with oil which serves to prevent oxidation of the mercury and to lubricate and to form a dashpot for the plunger.

Referring to the diagram of connections, Fig. 217, the solenoid A is connected in shunt across the generator terminals, so that the current flowing through it and the magnetic effect exerted by it are always proportional to the voltage at the generator terminals. Plunger C floats upon the mercury and if there were no other forces acting upon it than the buoyancy of the mercury and the force of gravity, it would always remain at a certain height. However, since the plunger is made of magnetic material, it is acted upon by the solenoid, the effect being to draw it farther out of the mercury the greater the current flowing through coil A. The resistance wire wound on plunger C is connected in series with the shunt field winding, but that portion of the resistance wire submerged in the mercury, of course, is cut out of circuit. Therefore, the greater the current flowing through solenoid A, the greater will be the outside resistance in circuit with the shunt field. The field current enters the regulator through binding post I, flows through the mercury in the inner well to pin D, thence through the resistance wire, into the mercury in the outer well, whence it flows into the frame or ground.

In this regulator provisions were made to compensate for the effects of temperature variations upon its operation. An increase in temperature would increase the resistance of the solenoid, thereby reducing the magnetic effect corresponding to a certain generator voltage. Consequently, more of the resistance wire would be submerged in the mercury and the battery would be charging at a higher rate than intended. Similarly, at very low temperatures, the battery would charge at a lower rate than intended. This is prevented by winding over the solenoid and connecting in series therewith a resistance wire of special material having a negative temperature coefficient (*i. e.*, whose electrical resistance decreases with an increase in temperature). In this way the total resistance in

GENERATOR CONTROL

the solenoid circuit remains the same irrespective of temperature changes.

Magnetic Vibrator Regulation—Several makers of electric equipment employ magnetic vibrators similar to coil vibrators for regulating the voltage or output of their generators. The

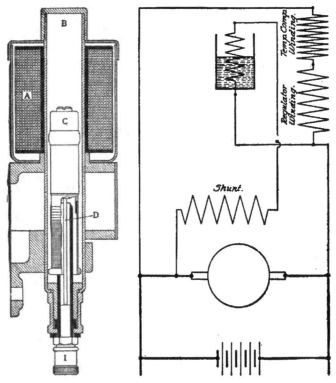

FIG. 216—DELCO MERCURY TUBE REGULATOR

FIG. 217—DIAGRAM OF CONNECTIONS OF DELCO REGULATOR SYSTEM

rate of vibration of such a vibrator depends upon the amount of current flowing through its coil—the greater the current the higher the rate of vibration. When such a vibrator is included in a field circuit the current flowing through the latter will be a pulsating one. Now, a field circuit generally has considerable self-induction, and the amount of current that flows through it will be the smaller the more rapid the pulsations, other things being equal.

It would hardly be feasible to connect the vibrator directly in circuit with the main shunt winding, as then the field excitation would momentarily drop down to nothing, and trouble would also be likely from the arcing which would accompany the breaking of a circuit of so much inductance. For that reason the vibrator is incorporated either in a supplementary shunt field (as in the Jesco system), or is connected across the terminals of a resistance unit in series with the shunt field.

A vibrator of the type employed for this purpose, the Bijur, is illustrated in Fig. 218. A is the iron core which is surrounded by a magnet winding H. Keeper G is pivoted at

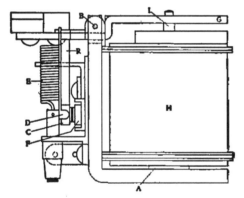

FIG. 218—BIJUR VIBRATOR REGULATOR

B and is held away from the end of the core by a coiled tension spring E. A brass stop pin I prevents the armature G from coming in contact with the end of the iron core. The contacts of the vibrator are shown at C and F. They are mounted on vibrating reeds, arranged at right angles to each other. Contact C and its reed R are mounted on armature G, and stop pins D limit the lateral movement of this contact. Contact F and its reed are mounted on an arm, as shown.

There is considerable tendency to arc when the contact is broken, and when magnetic vibrators were first applied to this purpose it was not uncommon for them to "freeze," that is, fuse together. In the regular operation of a vibrator, small particles of metal are constantly transferred from the positive to the negative contact, forming a crater on the positive and

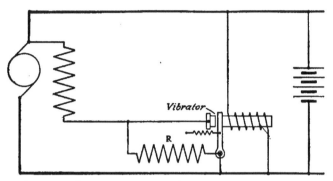

Fig. 219—Circuit Diagram of Constant Voltage System

a cone on the negative one. The roughness of the contact surfaces thus formed has a tendency to aggravate the arcing, and after this has been going on for some time the contacts are apt to stick together. The reed support for the contacts is intended to prevent trouble from this source. Owing to the vibration of the gasoline motor, the contacts will oscillate laterally between their stops, and in consequence the contacting surfaces will be continually shifting. The wiping surfaces will always keep clean and bright, it is claimed.

In Fig. 219 is shown the manner of connecting the vibrator in circuit. The generator is shunt wound and an outside resistance unit R is included in the shunt circuit. This resistance, however, is short circuited by the contacts of the vibrator (here shown in diagram) when these contacts are together, so

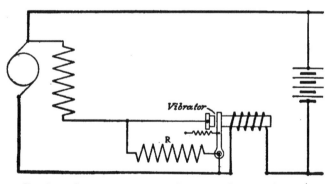

Fig. 220—Circuit Diagram of Constant Current System

that then the field current flows only through the shunt winding. When the vibrator contacts separate the resistance is included in the field circuit, and the field current is thereby reduced. As long as the vibrator is operating, the field current will constantly play between upper and lower limits, and its average value will depend upon the rate of vibration.

In Fig. 219, which illustrates the Bijur connections, the coil of the electromagnet is connected across the terminals of the generator. Hence the current through the coil will vary with the electromotive force developed by the generator. Any slight increase in generator *e. m. f.*, will cause a greater current to pass through the coil of the vibrator magnet, the

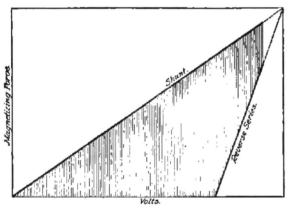

FIG. 221—DIAGRAM SHOWING RISE OF MAGNETIZING FORCE OF SHUNT WINDING AND DEMAGNETIZING FORCE OF SERIES WINDING WITH INCREASING VOLTAGE

vibrator will vibrate more rapidly and reduce the current flowing in the field winding, thus tending to keep the electromotive force developed constant. This, therefore, is an example of constant-voltage regulation.

A similar regulating system, but arranged for constant-current regulation, is used on the Ward Leonard electric system, and is illustrated in Fig. 220. We have here also a shunt wound generator with an outside resistance unit which is alternately cut into and out of the field circuit by a magnetic vibrator. But in this case the coil of the vibrator magnet, instead of being connected across the generator terminals, is connected in the main circuit. Therefore, the whole of the current delivered by the generator flows through it, and if this current increases the vibrator vibrates at a higher rate and cuts down the field current, thus tending to keep the current output of the generator constant. With both of these systems of control the charging rate can be adjusted by varying the tension on the vibrator spring.

Differentially Wound Generator—Undoubtedly the simplest system of control is by the use of a reverse series field winding. It does not involve any wearing or moving parts and requires no adjustments. The principle of action is exceedingly simple. At the speed at which the battery is connected to the generator by the automatic switch, the generator voltage only slightly exceeds the battery voltage and a small current will flow into the battery, which, by flowing through the reverse series field winding, produces a slight demagnetizing force. Any further increase in speed will increase the voltage and therefore send a stronger current into the battery. Now, the strength of both the shunt or direct field and of the series or reverse field increases with an increase in voltage, but the proportional increase in the strength of the series field is much greater. This is illustrated diagrammatically in Fig. 221. It is here assumed that a three-cell battery is connected to the generator, which begins to charge at 6 volts. Then the magnetizing force of the shunt winding is directly proportional to the generator voltage, whereas the demagnetizing force of the series winding is directly proportional to the excess in voltage over six. It will be seen that the line representing the demagnetizing force of the series winding rises much quicker than that representing the magnetizing force of the shunt winding, and if these lines were prolonged as shown by dotted lines, they would intersect. In other words, the direct magnetizing force of the shunt winding and the demagnetizing force of the series winding would be equal. But this cannot occur, because with two equal and opposite magnetizing forces there would be no magnetic flux, and no electromotive force would be induced in the armature. The voltage approaches the closer to that corresponding to the point of intersection the higher the speed, but it never quite reaches it.

This form of control gives neither a constant voltage nor a constant current characteristic, the current output rising rather rapidly to a maximum and then dropping off gradually, this latter effect being due chiefly to cross magnetization or shifting of the axis of commutation. Fig. 222 shows the current output of a generator with reverse series control, as dependent upon the speed of rotation.

Rushmore Ballast Coil Regulation—A modification of the reverse series system of control is employed by Rushmore.

GENERATOR CONTROL

One objection that may be urged against the plain differential generator is that the energy expended in the reverse series coil is wasted, at least at moderately low-charging speeds. The speed-current curve of such a generator is substantially like that shown in Fig. 222 and it would certainly be preferable to have the current attain substantially its full or maximum value at a lower speed and remain constant at higher speeds. Rushmore attained this end with very little complication. He made use of a peculiar property of iron wire—viz., that its resistance to the passage of electric current increases abruptly beyond a certain temperature. This is illustrated in Fig. 223. With an ordinary wire, say German silver, the amperes passed

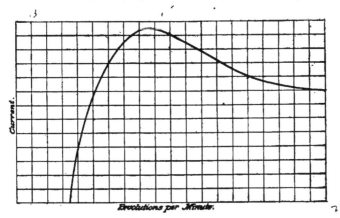

Fig. 222—Speed-Current Characteristic of Generator with Reverse Series Regulation

increase in direct proportion to the voltage applied, as shown by line B. With an iron wire of the proper dimensions, on the other hand, after the current has attained a certain value, a large increase in voltage produces practically no further increase in current, as shown by curve A. This property is made use of for regulating purposes in the following manner (Fig. 224):

The generator has both a main shunt field winding and a reverse series winding, the latter being called a bucking coil by Rushmore. Shunted across this series winding is a length of iron wire in the form of a ballast coil. Series winding and ballast coil are therefore connected in parallel and divide the main current between them in inverse proportion to their respective resistance; that is, the one having the most resistance takes the least current, and *vice versa*. Now, when the iron wire of the ballast coil is cool it has much less resistance than the series winding and takes by far the greater part of the current, so that the reverse series winding has little effect. But as the current increases the ballast coil heats up and its resistance quickly rises, so that most of the current delivered by the generator passes through the series coil and exerts a relatively strong demagnetizing effect. The result is that from the point where the generator begins to charge the battery the

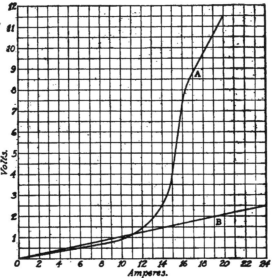

FIG. 223—DIAGRAM SHOWING ABRUPT INCREASE OF ELECTRIC RESISTANCE OF IRON WIRE AT CRITICAL TEMPERATURE

charging current increases rapidly with an increase in speed until it attains nearly its full value, and then, the critical temperature of the iron being reached, the current remains almost constant for any further increase in speed. The temperature at which the great increase in the resistance of iron occurs is just below red heat.

Another method of regulation by varying the magnetic flux through the armature consists in the provision of a centrifugal governor which draws the armature lengthwise part way out of the armature tunnel in proportion to the speed.

GENERATOR CONTROL

Magnetic Shunt—A further method of decreasing the magnetic flux as the output increases is to provide a magnetic shunt which forms a bypath for the magnetic lines of force past the armature. This plan is applicable only to certain forms of field frames. Fig. 225 illustrates the Wells generator which is regulated on this principle. It is of the two-pole type, with a single field coil located above the armature. The magnetic shunt A is hinged to one field pole, and is magnetically attracted by the other when there is current flowing through the field coil, the force of attraction, of course, increasing with the field strength. Upon an increase in speed, the armature voltage, and hence the field strength, will increase, which will result in the magnetic shunt being drawn closer to the field pole against the pressure of coiled spring B. This will shunt an increased proportion of the magnetic flux around the armature and thus keep down the armature voltage.

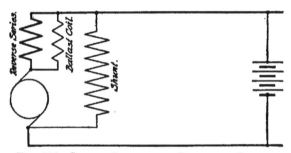

FIG. 224—CIRCUIT DIAGRAM OF RUSHMORE SYSTEM

Regulation Through Armature Reaction—It is possible to so design a shunt dynamo that it is more or less self-regulating. As has been pointed out previously, the voltage induced in the armature is proportional to three factors, viz., the speed of revolution, the magnetic flux and the number of armature conductors. In order to obtain a certain voltage the product of these three factors must have a certain value. But the designer is at liberty to vary the individual factors, as long as the product remains constant. For a self-regulating machine a comparatively small flux and a large number of armature conductors would be chosen. A small flux means a smaller cross-section of the magnetic circuit. It has repeatedly been stated in explaining the principle of previously described systems that with an increase in field current the magnetic flux increases. While this holds true for all degrees of magnetization, beyond the point of saturation a large increase in field current is attended by only a slight increase in magnetic flux. Practically speaking, it is then impossible to further increase the flux. This is one thing that tends to keep down the voltage during a rise in speed. Another is the considerable drop in voltage in an armature having a relatively large number of conductors, and the third and probably the most important cause tending toward self-regulation is armature cross-magnetization, or armature reaction. If we suppose that all parts of the magnetic circuit are already saturated when there is practically no current flowing in the armature, then a heavy armature current will not add to the flux through the armature, but it will shift the direction of the flux, and since the commutator brushes remain stationary they are no longer in the most favorable position for commutation. Looking at the matter in another light, the brushes are virtually shifted relative to the line of commutation, whereby the effective number of armature conductors is reduced and the voltage consequently kept down.

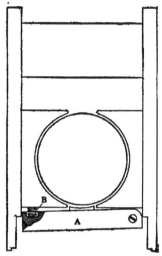

FIG. 224—WELLS GENERATOR WITH MAGNETIC SHUNT

Third Brush Regulation—A variation of the system of regulation just described, also depending on the principle of armature cross magnetization and the consequent shifting of the line of commutation, is that known as the third brush or interbrush method of regulation. The diagram, Fig. 226, explains this system. A shunt field winding is used and is connected between one of the main brushes and an extra brush located between the two main brushes. When the armature is carrying no appreciable current the main brushes bear on those commutator bars between which

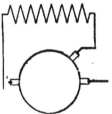

FIG. 226—THIRD BRUSH METHOD OF REGULATION

there is a maximum difference of potential. This is diagrammatically shown in Fig. 227. The curve represents the variation of potential around the commutator, and it will be seen that the two main brushes MB are located in the positions of maximum positive and maximum negative potential respectively, while the intermediate brush IB is located in an intermediate position. If now a load is put on the generator so that a heavy current flows through the armature conductors, the magnetic field will be distorted and the line of commutation rotated through a certain angle in the direction of armature rotation, as already explained. The conditions of commutation are then modified as represented by the brushes shown in outline. It will be noticed that, owing to the form of the curves, the potential differ-

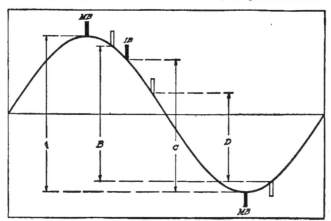

FIG. 227—DIAGRAM ILLUSTRATING VARIATION OF FIELD STRENGTH WITH GENERATOR OUTPUT IN THIRD-BRUSH REGULATION

ence between the main brushes is only slightly reduced, from A to B, whereas the potential difference between the intermediate brush and the main brush on the right (between which the shunt field coil is connected) is very materially reduced, from C to D. Therefore, as the speed of the generator increases and with it the charging current, the field strength is reduced so as to limit the current output. The effect may be accentuated by placing the main brushes slightly behind the axis of commutation for no load (to the left in Fig 227). Actually in a charging generator with third-brush regulation the charging current reaches a maximum at a certain speed and then decreases as the speed increases further. By the simple expedient of varying the position of the intermediate brush the speed-current characteristic of the generator can be changed.

The third brush regulating system has come into quite extensive use as it is very simple and seems to meet the requirements satisfactorily. Fig 228, based on information issued by the Dayton Electrical Laboratories Company, shows the variation in current output with speed of a generator with third brush regulation, as well as the variation of the field current.

Another method under the same heading consists in mounting the field frame on trunnions. As the armature revolves there is a certain magnetic drag between armature and field,

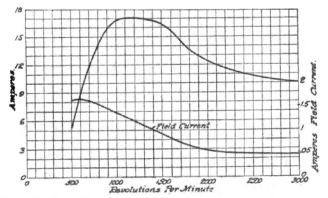

FIG. 228—SPEED-CURRENT AND SPEED-FIELD CURRENT CHARACTERISTICS OF GENERATOR WITH THIRD-BRUSH REGULATION

which increases in proportion to the current flowing in the armature. The field frame tends to follow the armature in its motion, but is restrained by a spring. The spring will yield an amount proportional to the armature current, and since the brush rigging is stationary the brushes will be moved from the line of most favorable commutation in proportion to the load on the generator. This involves a reduction of the number of effective armature conductors and consequent regulation of the output.

Cam-Operated Regulator—An interesting method of voltage control has been introduced by the Westinghouse Electric & Mfg. Co., which in the past has used the reverse series and the third brush methods. This consists in the use of a vibrator of peculiar form for introducing a resistance into the shunt field circuit of the generator and cutting it out periodically. The regulator is illustrated in Fig. 229. The vibrating blade or reed is acted upon by a cam, or rather, an eccentric, which is keyed to the armature shaft. This eccentric has a very small eccentricity, 0.008 inch in one particular design. What corresponds to the stationary contact in an ordinary vibrator is flexibly supported and is acted upon by an electro magnet carrying a winding connected across the armature terminals. As the eccentric revolves the vibrating contact point moves up and down, and it remains in contact with the other point for a shorter or longer fraction of the time of each

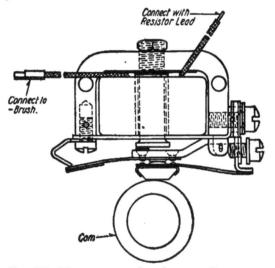

FIG. 229—WESTINGHOUSE CAM-OPERATED REGULATOR

revolution, according to the position of that point, which is controlled by the electro-magnet. As the generator voltage increases, the magnet becomes stronger and draws the upper contact farther away from the lower. The result is that contact will last for a shorter time. Fig. 230 shows the electrical connections. It will be seen that when the contacts are closed the field resistance is cut out of circuit and when they are open it is in circuit. Therefore, as the contacts are moved farther apart and in consequence come in contact for a shorter period of time, the effect of the field resistance is increased, *i. e.*, the tendency is to diminish the generator voltage and thus to neutralize the effect of the increase in speed.

GENERATOR CONTROL

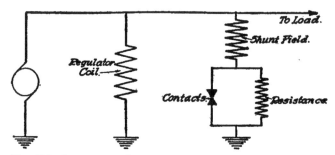

FIG. 230—DIAGRAM OF CONNECTIONS OF CAM-REGULATOR SYSTEM

This method of voltage regulation has been employed on motor trucks and motorcycles where, for certain reasons, it was considered undesirable to carry a battery. The absence of a battery, of course, renders the problem of voltage control for electric lighting much more difficult, as a battery floating on the line has a powerful steadying effect. Very good regulation is obtained with this system as may be seen from Fig 231, which gives the variation of the voltage with engine speed for three different loads, viz., headlamp, sidecar lamp and tail lamp all on (full line); head lamp and tail lamp only (dashed line), and tail lamp alone (dash-dotted line). It will be seen that the voltage is maintained very nearly constant throughout a speed range extending from about 1000 r.p.m. to about 3000 r.p.m. of the engine. The generator was driven at twice crankshaft speed. The little machine was of cylindrical form, 3 7/16 inches in diameter by 6¼ inches long over all and weighed only 7¼ lb.

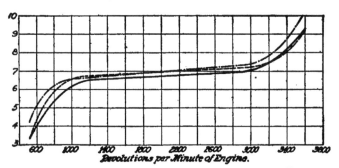

FIG. 231—SPEED-VOLTAGE CHARACTERISTIC AT DIFFERENT LOADS OF CAM-CONTROLLED GENERATOR

GENERATOR CONTROL

Effect of Temperature on Generator Output—The output of an electric generator has a natural tendency to decrease as the temperature of the generator rises. This is due to the fact that the resistance of all of the wires increases with a rise in temperature, and in consequence there is a greater voltage drop in the armature windings, and less current flows through the field windings for a certain voltage impressed upon them. This is a rather fortunate condition, for during the winter, when the atmospheric temperature is low, more electrical energy is required for both starting and lighting. By careful design the effect of temperature variation upon output can be farther increased. Of course, after the generator has been

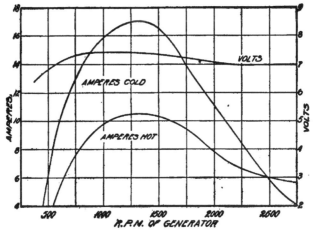

FIG. 232—EFFECT OF FRAME TEMPERATURE ON GENERATOR OUTPUT

running for an extended period its temperature will be much the same, regardless of atmospheric temperature, but the charging rate may be very considerably greater while the machine is first started from cold. This feature of charging generator design has been specially investigated by the Westinghouse Co. and Figs. 232 and 233 show the characteristics of a generator with third brush method of control developed by that concern. With this system the drain on the battery due to short trips, or runs involving frequent starts, is met by larger current output from the generator. More current can be taken from a generator with this method of design than it would be capable of withstanding continuously without overheating. The very rise of temperature that would tend to endanger the generator automatically protects it by cutting down its output to a safe value.

Thermostatic Control of Generator Output—The increased demand for electric energy in winter driving has led the Remy Electric Co. to provide a thermostat to increase the charging rate in cold weather. This device cuts a resistance in the field circuit at high temperatures, thus decreasing the output. The thermostat proper is of the dissimilar metals type. Referring to Fig. 234, steel punchings A support the thermometal blade B with silver contact points C, and the resistance unit E, A and B are riveted together through block D in such a manner that A is insulated from B. The resistance

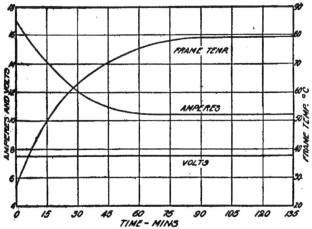

Fig. 233—Rise in Frame Temperature and Drop in Output after Starting from Cold

unit is formed by winding resistance wire around a heavy piece of mica.

The thermo blade B is composed of a spring brass strip welded to a strip of nickel steel. The brass strip is on the lower side of blade B and tension has been placed on the spring brass through the medium of the adjusting nut F so that the points C are always in good contact at temperatures below 150 deg. Fahr. This adjustment is made at the factory and is permanent.

If the temperature of the air about the thermostat exceeds 150 deg. Fahr. the blade is warped upward owing to the greater coefficient of heat expansion of the brass as compared

284 GENERATOR CONTROL

with the nickel steel, thus separating contact points C. Conversely, when the temperature falls below 150 deg. Fahr. the spring brass strip pulls down and the points C again make contact.

The generator to which this thermostatic regulator is applied is controlled by the third brush method, as shown by the diagram of connections, Fig. 235. With the field resistance in circuit the maximum output is 14-15 amperes, with the resistance out of circuit it is 20-22 amperes.

With a thermostatic switch as here described, if a driver starts out on a summer day, the generator output will be about 21 amperes. After several miles have been run at normal speed the thermostat will open and the output will drop to about 15 amperes. In the meantime the energy con-

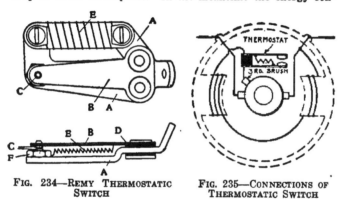

FIG. 234—REMY THERMOSTATIC SWITCH

FIG. 235—CONNECTIONS OF THERMOSTATIC SWITCH

sumed in cranking the engine has been returned to the battery. On the other hand, during a very cold day the thermostat will never open.

Care of Automatic Regulators—With a governor clutch, wear of the friction facing will lower the pressure of the spring when the clutch is in engagement. Hence, less centrifugal force on the governor weights will be required to disengage the clutch and the latter will then disengage at a lower speed. The result of this will be that the generator will run at lower speed and will generate less current than formerly. Therefore, if the battery will not remain charged and there is no other way to account for it, it may be that the clutch facing has worn down, and in order to compensate for this wear the spring pressure must be increased by means of the adjusting nut.

With a system having a variable resistance in the field circuit controlled by the main current, if one of the battery connections should come loose or one of the wires leading to the batteries should break, then the magnet or solenoid controlling the resistance in the field circuit would become practically ineffective. Suppose, for instance, that only the side and tail lamps were turned on. The solenoid would begin to cut resistance into the field circuit only when the current in the main circuit had attained the normal charging rate, and long before this the lamps would be burned out. To obviate such trouble a fuse is inserted in the field circuit, which, when the generator voltage materially exceeds the normal value, burns out and thus protects the system. Hence a blown field fuse with this system indicates a loose or broken battery connection, and whenever such a fuse is replaced the cause of its blowing must be looked for and remedied.

The result of a loose or broken battery connection is about the same in a system employing a magnetic vibrator regulator controlled by the main current. In this case the vibrator would not operate until a dangerously high voltage had been attained. Fortunately, battery connections now are made in such a secure manner that the chance of their breaking or becoming detached is rather remote.

Regulators with moving parts must be kept clean and in a free-moving condition. The contacts of vibrators must also be kept clean and smooth, and in case they should become rough they can be dressed up with a fine, smooth file. With the latter type of regulators, arcing at the contact points is the most serious source of trouble and some preventive measure is applied by almost every maker. The method of mounting the contacts on reeds arranged at right angles to each other has already been referred to. One manufacturer uses a contact disk instead of the stationary contact point, which he rotates positively so that the point of contact on the disk is changed continually. Of course, a condenser may be shunted across the vibrator, the same as in an ignition system, but we do not know of any case where this has been done. A substitute for a condenser, known as a resistance condenser and consisting of a spool of fine wire wound doubled up so as to make the coil inductionless, is sometimes used.

CHAPTER XXIV

Battery Switches—Charge Indicators

The armature of a battery charging generator has exceedingly little resistance, and if the generator, while at rest, were electrically connected to the battery, its armature would practically short-circuit the latter. This is what actually takes place in a starting motor, and the armature current in such a motor attains an exceedingly high value for a moment, but the motor immediately begins to revolve and generate a back e.m.f., which results in a rapid reduction of the armature current. In the case of a generator, the gear ratio of its connection to the engine is often such that it would not be able to turn if the battery circuit were left closed when the engine was stopped, and the result would be that the battery would become discharged in a very short time, and most likely the generator winding would be injured. Therefore, some means must be provided for disconnecting the battery from the generator when the engine is stopped.

Hand Battery Switch—The simplest means consists in a hand switch mechanically interconnected with the ignition switch in such a way that when ignition is cut off to stop the engine the electrical connection between the generator and battery is also interrupted. This sort of switch is used only in connection with systems in which a single electric machine serves both as charging generator and starting motor. With a hand switch the battery will not be disconnected from the dynamo when, because of an unusual reduction in the speed of the latter, its voltage falls below that of the battery. In that case the battery will discharge through the dynamo and the latter will act as motor, assisting the gasoline engine to carry its load. This gives the rather important advantage of a practically unstallable engine, which increases the element of safety in a gasoline automobile.

A typical hand battery switch is that of the Entz system. The switch is of substantial construction and is mounted on the dashboard. Its operating parts are enclosed in a cast

BATTERY SWITCHES

aluminum housing, from which the switch handle extends at the top. It may first be pointed out that the Entz system. comprises a dynamotor with differential field winding, *i.e.*, both a shunt winding and a reverse series winding; also that, as represented in the diagram, a magneto is used for ignition and, therefore, when the ignition switch is closed the magneto is short-circuited and ignition is "off," whereas when the switch is open ignition is "on." The switch has three positions, as shown in Fig. 236. In the normal running position, 3, the generator switch is closed and the magneto switch is open, so that the ignition system can operate and the battery charge; in position 2, or intermediate position, both the gen-

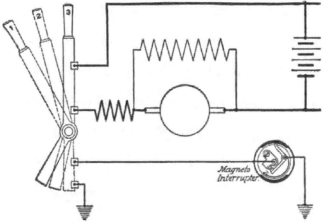

FIG. 236—HAND BATTERY SWITCH (ENTZ)

erator switch and the ignition switch are open, so that no current can now flow from the generator to the battery, or vice versa, but the ignition system can operate. In position 1 the generator switch is open but the magneto switch is closed, so that ignition is shut off and the generator is disconnected from the battery. In this, the "off" position, the switch handle can be locked by means of a tumbler lock, to prevent unauthorized persons from starting the car. This system offers the advantage over most others that if the battery is fully charged the operator can run the car with the switch in the intermediate position, in which no current is generated and none therefore is wasted in unnecessarily heating up the electrolyte. To start the engine, the operator moves the switch lever from position 1 to position 3, and ordinarily he leaves it there as long as the engine is running, the dynamo beginning to charge the battery as soon as the engine has attained a certain speed.

Centrifugal Switch—The great majority of lighting and starting systems, however, employ an automatic switch to "cut in" the battery when the generator voltage, on increasing speed, has attained a value high enough to overcome the battery voltage and send a charging current into the battery, and "cut out" the battery when the generator voltage, on decreasing speed, becomes too low for charging. These automatic switches act either by centrifugal force or by electromagnetic attraction. One of the simplest of centrifugal switches, that used on the Mira, an English system, is illus-

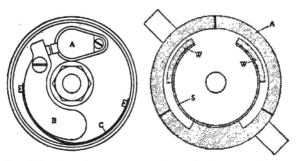

FIG. 237—MIRA CENTRIFUGAL SWITCH

FIG. 238—HOLTZER-CABOT CENTRIFUGAL SWITCH

trated in Fig. 237. The device is in some respects similar to a magneto interrupter, being carried on the end of the revolving armature shaft. Since the parts forming the switch revolve with the armature, the current must be carried to them by means of a collector ring (or disk) and brush, and this part, designated in Fig. 237 by A, is combined with a "stationary" contact point. The movable contact is carried on a centrifugal weight, B, and is held away from the stationary contact by the flat spring, C. The weight being pivoted unsymmetrically, when a certain speed of rotation is attained the centrifugal force on the weight causes the contacts to come together and the battery circuit to be closed.

Another very simple form of centrifugal switch is the Holtzer-Cabot, represented in Fig. 238. It consists of a cylindrical housing, A, made up of three equal sectors insulated

BATTERY SWITCHES

from each other. One of these sectors carries a pair of centrifugal weights, WW, supported on a flat steel spring, S. As the housing revolves the weights fly out and make electrical contact with the two remaining sectors, thus connecting all three sectors together. A pair of oppositely arranged brushes bear on the outside of the cylindrical housing, and when the centrifugal weights contact with the sectors the current can flow right through from brush to brush, which it can not do when the weights are out of contact with the sectors.

A third form of centrifugal switch comprises a regular centrifugal governor with a sliding sleeve or thimble over the end of the armature shaft, the thimble carrying a contact

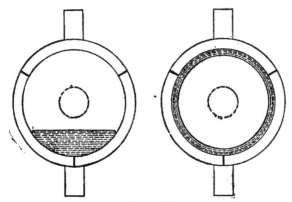

FIG. 239—MERCURY TYPE CENTRIFUGAL SWITCH

button at the end which is pressed by the centrifugal force on the governor weights against a stationary contact button in line with the armature shaft.

Mercury Type Battery Switch—In some instances battery switches in which centrifugal force and friction act on a quantity of mercury in a revolving cylindrical chamber are made use of. One such switch is illustrated in diagram in Fig. 239. The cylindrical wall of the mercury chamber consists of three equal sectors which are insulated from each other. Two brushes bear on the outer surface of the cylindrical wall at diametrically opposite points. Evidently, since each of the sectors measures less than 120 degrees, no current can normally pass through them from one brush to the other. Inside the chamber, which forms a closed vessel, there is a quantity of mercury, which, when the device is at rest, lies at the bottom. When the chamber begins to revolve, friction between the mercury and the wall of the chamber tends to carry the mercury along in the rotary movement, while the force of gravity tends to keep it at the bottom. The result is that the mercury spreads out over the cylindrical surface until it covers more than one full sector at the same time, and then there is an unbroken path for the current through the device. At high speed the centrifugal force keeps the mercury distributed in an even film or layer over the whole internal surface of the chamber. Friction no longer plays a part, because the mercury revolves with the case without slippage. If now the speed decreases, a point will be reached when the centrifugal force is no longer sufficient to keep the mercury applied against the wall of the chamber, so that it falls to the bottom and interrupts the circuit. Hence, the closing speed of such a switch depends upon the friction between the mercury and the case and the opening speed upon the centrifugal force upon the mercury. This permits of so constructing the switch that it will not open at the same speed at which it closes but at a somewhat lower speed, which offers material advantages, as will be explained in connection with electromagnetic switches.

Electromagnetic Battery Switch—Most battery charging systems employ an electromagnetic switch to open and close the battery circuit at the proper moments. These switches are compact, simple and comparatively reliable in operation. One of the reasons for their prevalence is probably that the men who design electrical equipment are more familiar with this type of mechanism than the others available for the purpose. An electromagnetic switch is illustrated partly in diagram, in Fig. 240. The electromagnet, A, has both a shunt winding and a series winding. It is provided with an armature, B, which is normally forced away from the magnet pole by a spring, C. As soon as the generator begins to generate, a current begins to flow through the shunt or fine wire winding of the switch magnet. This current continues to grow as the speed of the generator (and with it the voltage) increases, and when the voltage is sufficiently high to send a charging current through the battery the electromagnet attracts its armature, B, which carries the movable contact, D, and brings the latter up against the stationary contact, F, thus closing the battery circuit. The path of the charging current can now be easily traced out. It will be seen that this current passes through the series or heavy wire winding on the switch magnet, thus further strengthening the magnet and holding the switch closed the more firmly. This, however, is only incidental, the real object of the series coil being to aid in opening the switch.

As the voltage of the generator decreases by reason of a decrease in speed, the magnetizing force of the shunt winding on the electromagnet varies in direct proportion to the e.m.f. When the generator voltage has come down to the same value as the battery voltage the switch should open, but the magnetizing force of the shunt winding is then still nearly as strong as when charging at the normal rate, and, since it does not require nearly as great a magnetic force to hold the armature on the core as to draw it there, if only a shunt winding were used on the electromagnet the switch might not open when the generator voltage fell below the battery voltage. This

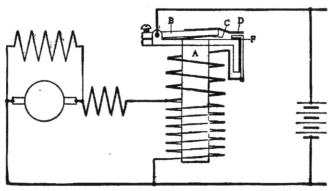

FIG. 240—ELECTRO-MAGNETIC SWITCH, PARTLY IN DIAGRAM

is where the advantage of the series coil comes in. When the generator voltage has come down to the value of the battery voltage no current flows through the series coil any more, hence the latter has lost all of its magnetizing force, and, if the generator voltage drops still further, then a reverse current is sent by the battery through the series coil, creating a demagnetizing force. This latter will increase in value until the effective magnetizing force (the difference between that due to the shunt winding and that due to the series winding) is so low that the spring forces the armature away from the core and opens the switch.

Fig. 241 shows the Delco cut-out switch, which is of the electromagnetic type. A represents the switch contacts (which are double); B, the cut-out spring; C, the armature, and D, the core of the magnet. This cut-out is so adjusted that when the points are in contact there is an air gap of about 1/32 inch between the core and the armature. By means of a lever F with ball handle E the driver can close the circuit manually and thus use the cut-out also as a starting switch.

Centrifugal and Magnetic Switches Compared—A centrifugal battery switch differs in its action from an electromagnetic one in several ways. With a centrifugal switch, if the generator for any reason fails to "pick up," this does not matter, for when the proper speed is attained the battery is connected to the generator and battery current flows through the field coils, which causes the field to immediately assume its full strength and starts the generation of current without fail—unless the machine is in an inoperative condition. On the contrary, with an electromagnetic switch, if the generator fails to pick up there is no current to operate the switch and the battery will not begin to charge. This, of course, can be remedied by closing the battery switch by hand, whereupon battery current will excite the field of the generator. In a differentially wound machine, if battery current is sent through the field coils, the series and shunt fields act in the same direction, instead of opposing each other, and the field therefore almost instantly attains an unusual strength. In this connection it may be well to caution the operator never to close the battery switch by hand when the engine is not running, as battery current will hold it closed until the battery is practically discharged, which will not be very long.

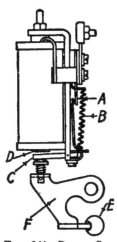

FIG. 241—DELCO CUT-OUT SWITCH

Another difference between centrifugal and electromagnetic battery switches is that with the former it does not matter which way the battery is connected to the generator, provided the switch is so designed that it closes at a speed slightly below that at which the generator voltage equals the battery voltage. Then, immediately upon closing the switch, battery current will flow through the generator field coil, and if the battery were incorrectly connected the polarity of the field would be instantly changed and the generator e.m.f. would then be in the right direction for charging.

Electromagnetic switches are always so designed that they open the circuit at a somewhat lower speed than that at which the circuit is closed. If opening and closing speeds were substantially alike, then, if the generator happened to be running at about this speed for some time, the switch would be continually opening and closing, and the resulting arcing would be injurious to the contact points. There is a natural tendency for the switch to open at a lower speed than the closing speed, for the reason that it requires much less magnetizing force to hold the keeper in position on the core than to draw it there from its position of rest.

Location of Switch—Battery switches operated by centrifugal force are always located on the generator, for obvious reasons. Switches of the electromagnetic type also are often located at the generator, on account of the simplicity of the outside wiring connections resulting therefrom. For instance, in a generator system with electromagnetic switch and current or voltage regulation by means of a field resistance and magnetic vibrator, if switch, resistance and vibrator are located in or on the generator housing, only two cables need to lead away from generator—those leading to battery. However, there is one

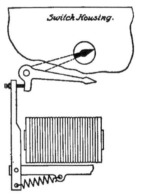

Fig. 242—U. S. L. Charging Indicator

reason which sometimes leads designers of electric equipments to arrange the electromagnetic battery switch for connection to the dashboard. Since the battery is charging whenever the switch is closed, the switch itself forms a charging indicator. The position of the switch, however, is not clearly visible from a distance, and for this reason it is usually made to actuate some indicating device either in the form of a pointer or a sign appearing behind a window in the switch housing. A pointer is used in the U. S. L. system, as shown in Fig. 242. It is made in the form of a bell crank with unequal arms. A setscrew in the end of the switch lever presses against the short arm of the bell crank when the switch is closed and raises the tip of the pointer to a position

CHARGE INDICATORS

directly behind a small window in the housing. When the switch opens the pointer drops back by reason of its weight.

Charge Indicators—The movable sign type of charging indicator is based on exactly the same principle as the pointer type. A lever or bell crank actuated by the switch lever, or by the magnet core directly, carries two signs. When the switch is closed the sign "Charging" appears behind a window in the casing and when the switch is open the sign "Off" appears there.

With an indicator thus actuated from the battery switch it is not possible to indicate a discharge of current, for the switch has only two positions—open and closed. The switch is inserted in the line between the generator and the battery and is not affected by any current that may be drawn from the battery through a separate circuit. It is possible, however,

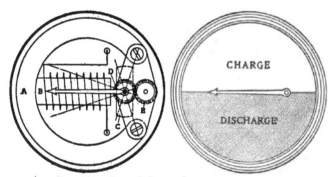

FIG. 243—GRAY & DAVIS CHARGING INDICATOR

to arrange a charging indicator so it will indicate both charging and discharging currents, and an instrument of this type (Fig. 243) is furnished with the Gray & Davis system. It comprises a permanent magnet, A, of true horseshoe form, which exactly fits into the round casing. Within the horseshoe there is a piece of soft iron, B, which is surrounded by a coil whose ends are brought out to binding posts on the back of the instrument. In the space between the poles of magnet, A, and the near end of piece, B, is located a light bar, C, of soft iron carried on a spindle which also carries the pointer moving over the dial plate of the instrument and a small spur pinion, D. The latter meshes with another similar pinion, E, on a spindle parallel to the first, to which is secured a small spiral spring (not shown) which holds the movement in the central or neutral position. Now, suppose that a current is flowing through the coil in such a direction that it makes the lower end of piece B a north pole. Then the armature tends to set itself in line with this north pole and the south pole of the permanent magnet, deflecting the pointer and indicating "Charge" or "Discharge," as the case may be. When the current flows through the coil in the opposite direction it makes the lower end of the soft iron piece a south pole and the soft iron bar then tends to set itself in line with this south pole and the north pole of the permanent magnet, thus deflecting the pointer in the opposite direction. As shown in the second view, the lens of the instrument is divided into two fields, one of which is colored. When no current flows the pointer is in line with the division between the two fields.

CHAPTER XXV

Lamps and Fittings

Law of Illumination—In the case of a free source of light, the intensity of illumination varies inversely as the square of the distance from the source. This is easily proven as follows: Suppose the source of light to be a point at the center of a hollow sphere of one foot radius. Then, since every point of the interior surface of the sphere is at the same distance from the source, the illumination is uniform over the whole surface. If we suppose that the source of light is of unit power, then the illumination on the inner surface of the sphere is of unit intensity. If, now, we surround the source of light with a hollow sphere of two feet radius, then the luminous flux emitted by it is distributed over an area four times as great as in the previous case (since surface area varies as the square of the radius) and the illumination of the surface is only one-fourth as strong as it was previously. Therefore, intensity of illumination varies inversely as the square of the distance from the source.

Unit of Light—The present unit of light is the international candle power which is based on the mean intensity of a group of incandescent electric lamps maintained by the U. S. Bureau of Standards at Washington. It is very nearly the same as the old English candle power. This is the light given by a candle made of spermaceti, a wax-like substance obtained from whale oil; the candle is $7/8$ inch in diameter, burns with a flame $1 3/4$ inches high and consumes 120 grains of spermaceti per hour. These candles were used for photometric measurements for a long time, but were finally given up because their luminous intensity varied as much as 30 per cent. The method employed in measuring light is based on the law of the variation of illumination with the distance from the source and upon the fact that a grease spot on a sheet of white paper appears light on a dark ground under transmitted light (if illuminated from the back) and dark on a light ground under reflected light (if illuminated from the front or the side of the observer). If the paper is illuminated equally strongly from both sides the grease spot is invisible. Hence, in measuring the strength of light the standard and the light to be measured are placed at a certain distance from each other on a base, with a movable screen in between. On this screen there is a central grease spot. The screen is shifted until the grease spot on it becomes invisible. The distances between the screen and the two lights, respectively, are measured or read off and the candle power of the light being measured is then expressed by the quotient of the square of the distance between screen and light to be measured, by the square of the distance between screen and standard candle.

The apparatus thus described constitutes the simplest form of photometer. Many different types of photometer have been developed during the past two or three decades and highly refined for quick and accurate work with different types of lamps.

Parabolic Reflector—On an automobile it is desirable to throw a strong beam of light in only one direction, and as the power of the lights is limited, the rays issuing from the source in all directions must be reflected in the direction in which illumination is desired. In electric lamps this is always accomplished by means of parabolic reflectors. A parabola is a plane figure having the peculiarity that the distance from any point in its bounding curve to a certain point called the focus is equal to the distance from the same point in the curve to a line called the directrix. In Fig. 244, which shows a parabola, F is the focus and OA the directrix. The distance FX is equal to the distance XB, and this holds true whereever X may be located on the curve. A parabolic reflector, of course, is a shell whose section forms a parabola.

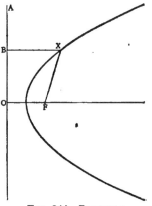

FIG. 244—PARABOLA

It can be seen from Fig. 245 that if a luminous source is placed at the focus of the parabola all of the rays emanating from it which strike the parabola are reflected parallel to the axis of the parabola. This is in accordance with the law of optics that the angle of incidence is equal to the angle of reflection. Theoretically, therefore, we get a parallel beam whose section is equal to the opening of the reflector. Also, since the rays are parallel, there is no diminution in the illumination with an increase in the distance from the source due to spreading or dispersion, as in the case of free light, but loss of intensity is caused by absorption of light by the atmosphere. In practical work, although the reflector may be a perfect paraboloid, the beam of light projected will not be a parallel beam, because the source of light is not a point, but a surface of considerable area, most of which is at some distance from the focus of the reflector. Some enlargement of the shaft of light is, of course, desirable, as the road in front of the vehicle should be lighted substantially uniformly over

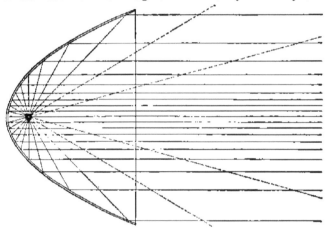

Fig. 245—Showing Parallel Reflection of Rays From a Point Source at the Focus of the Parabola

a width somewhat greater than that of the vehicle, besides which it is desirable that a less intense light be thrown on to the roadsides.

The farther the reflector extends in front of the source of light the greater the proportion of the total light emitted by the source which is caught by the reflector and condensed into a shaft. The remaining rays of the source, as indicated by dotted lines, are dispersed and furnish roadside illumination. Therefore, the efficiency of a parabolic reflector, so far as the shaft of light thrown ahead is concerned, is measured by the proportion of the solid angle subtended by the surface of the reflector at the focus to the solid angle represented by a whole sphere.

LAMPS AND FITTINGS

Definition of Terms—It has been customary to denote both the bulb and the fixture by the term "electric lamp," and, to obviate confusion and error due to this double use of the word, the Electrical Equipment Division of the S. A. E. Standards Committee adopted the following definitions:

Bulb—The detachable electric light-giving unit comprised of a filament and its glass envelope.

Lamp—The fixture for mounting and utilizing the light of the bulb.

An electric bulb (Fig. 246) consists of a filament of conducting material enclosed in a glass bulb from which the air has been exhausted. The ends of the filament are connected to short lengths of thin wire (the leading in wires) fused into a

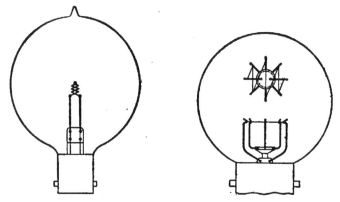

FIG. 246—TWO TYPES OF AUTOMOBILE BULBS

glass stem extending into the bulb at the base end, which in turn connect to metal parts on the base of the bulb. Originally carbon filaments were used exclusively. These were made from some carbonized fibrous material, usually amorphous cellulose. It may be pointed out that the object in exhausting the air from the bulb is to prevent combustion of the filament in the oxygen of the air when it is heated to redness. With no air present the filament cannot burn.

Development of Metal Filament—Carbon filament lamps consumed about 3.5 watts per candle power, and great efforts were made to reduce the energy consumption, either by improving the carbon filament or by using filaments of other materials, it being realized that the absolute efficiency of the lamp was very low, by far the greater part of the energy consumed being converted into heat and only a very small fraction into light. A great advance in gas lighting had been brought about by the introduction of the incandescent gas mantle invented by Dr. Auer von Welsbach. This consists of a fabric impregnated with certain metallic oxides, which is heated to incandescence by the gas flame and gives out more light than the gas flame itself would. Dr. Welsbach conceived the idea of using these same oxides in electric lamp filaments, and produced an osmium filament lamp, which was a considerable improvement over the carbon lamp. Black osmium powder was mixed with a fluid binder to form a coherent mass, which was extruded through a die into a filament of the desired thickness, and this was then baked in an electric furnace. Tantalum and tungsten filaments were made by similar processes. Later it was found possible to produce metallic tungsten in a sufficiently malleable form to draw it into thin wires, but owing to its extreme hardness diamond dies had to be used instead of the usual steel dies for wire drawing. Most of the present automobile lamps are tungsten wire filament lamps. The larger sizes, such as used for head lamps, consume about 1 watt per candle power, while the smaller lamps consume about $1\frac{1}{4}$ watts per candle power.

Some of the properties of tungsten lamps were brought out in a paper by Henry Schroeder on Electric Bulbs for Automobiles read before the Society of Automotive Engineers in 1916. Tungsten has a very high temperature coefficient and at the normal working temperature of the electric bulb the resistance is 12.5 times as great as when cold. On account of the variation in resistance of filaments at different temperatures, varying voltages do not produce proportional variations in current and energy consumption. The greater the increase in resistance with increase in voltage, the less will be the increase in amperes and watts. This gives the tungsten filament an inherent advantage over other filaments, such as tantalum, metalized carbon and carbon, which latter has even a negative temperature coefficient.

The candle power of a filament is very sensitive to changes in voltage, as slight changes in temperature produce great changes in candle power. On account of the lesser variation in the wattage consumption with voltage changes of the tungsten as compared with the other filaments, the tungsten filament varies less than other filaments in candle power with varying voltage. The characteristics of the tungsten bulb are given in Fig. 247, taken from Mr. Schroeder's paper. It will be seen that with increasing voltage the amperage increases mod-

LAMPS AND FITTINGS 301

erately, the wattage faster and the candle power still faster, while the watts per candle, which is an inverse measure of the efficiency, decreases.

Types of Filaments—An important point in automobile bulbs is to arrange the filament as compactly as possible so that it may approach a point source of light. In a parabolic reflector only the rays emanating from the focus are projected parallel to the reflector axis. Rays from points outside the focus are dispersed more or less, the more the greater the distance of the point from the focus, whereby the efficiency of the light is reduced. In a bulb for a non-focusing lamp, that is, a side or tail lamp, compactness of the filament is not

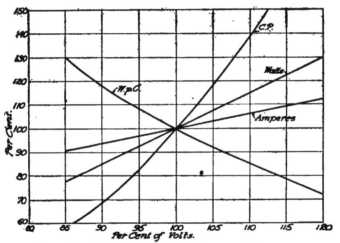

FIG. 247—CHARACTERISTICS OF TUNGSTEN FILAMENT

so essential and the form of filament shown at A in Fig. 248, known as the loop back filament, is then generally used. It is anchored at the middle to give it a more rigid support and thus enable it to better withstand vibration. There are several forms of high candle power bulbs. In one, shown at B, the filament forms a single compact coil whose axis coincides with the axis of the bulb, with a straight return portion in the center of the coil. An even more compact arrangement, consisting of a double coil, is shown at C. The filament shown at D consists of a fine coil strung between two supporting wires and that at E is arranged in V shape and carried by three wires. In still another form, shown in Fig. 246, the filament is strung in zig-zag form between a series of supporting wires in a plane perpendicular to the axis of the bulb.

Nitrogen-Filled Bulbs—Quite recently automobile type bulbs have been placed on the market, which, instead of being exhausted, are filled with nitrogen gas. Most gases react with tungsten at a high temperature, but exhaustive experiments showed that nitrogen is inert to tungsten under these conditions. Nitrogen-filled bulbs show a much higher current economy than vacuum bulbs. In the larger sizes of these bulbs for house lighting, etc., an economy of ½ watt per candle power has been obtained, but the efficiency of the small electric bulbs for vehicle lighting is not quite so high. Headlight bulbs of the nitrogen-filled type of 21 candle power consume from 15 to 17 watts and bulbs of 24 candle power from 18 to 21 watts, making an average economy of about ¾ watt per candle power. Some of the chief characteristics of nitrogen-filled bulbs, as compared with ordinary tungsten

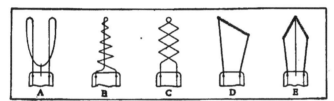

FIG. 248—TYPES OF FILAMENT

lamps, are as follows: The light is much whiter and approaches daylight much closer. The brilliancy of the filament is much greater, hence a filament producing a given total radiation of light is much more compact, and such a bulb, therefore, is much better suited to headlights. Furthermore, the characteristics of the nitrogen-filled bulb are more nearly constant throughout its life, that is, its candle power falls off less with age than that of regular tungsten bulbs. It is also claimed to have a much greater life.

Lamp Voltage—For several years after electric lighting was first introduced 6 volts was the common voltage. It was adopted partly because it was the standard ignition voltage, hence current for ignition and lighting could be taken from the same battery; and partly because the lower the voltage of a lamp of given type and candle power the greater the mechanical rigidity of the filaments. There is still one more advantage in a low voltage system, and that is, that for a given battery capacity the battery weighs less the lower the voltage and hence the smaller the number of cells employed. For instance, a 6-volt, 720-watt-hour battery of a certain make weighs 68 lbs. and an 18-volt battery of the same capacity 90 lbs. The chief disadvantage of such a low voltage is in connection with the starting motor. The current drawn by the latter on 6-volt systems attains enormous values, and unless exceedingly heavy cables are used there is a great loss of voltage in the conductors. To obviate this difficulty, multi-voltage systems were much used in former years. Either six, nine or twelve cells were employed, which were connectd all in series for starting and in two or more parallel groups for charging and lighting. This necessitated a rather complicated wiring and switching system, and it is now rarely used. On the other hand, it has been found possible to make bulbs for 12 volts which will show a satisfactory life, and a certain number of straight 12-volt systems have been developed. Twelve volts is the standard abroad, notwithstanding the fact that starting motors are seldom operated from the lighting batteries there. There are, therefore, at present two common voltages, viz., 6 and 12, though the former is used much more extensively. In the S. A. E. paper on Electric Bulbs already referred to it was stated that in 1915 eight million bulbs for three cell (6 volt) and one and a half million for six cell (12 volt) systems were used in the United States, and since then the 6-volt system has made further gains.

Candle Powers, Life and Efficiency of Bulbs—Bulbs for head lamps are made in 15 to 24 candle powers and bulbs for side and tail lamps in 2, 4 and 6 candle powers. Bulbs vary in size according to the candle power, but the bases of all are alike. All bulbs for outside lamps are spherical in shape, and the center of the filament coincides substantially with the center of the bulb, the filament in the larger bulbs being placed at a considerable distance from the base, in order that the latter may not shut out too great a proportion of the light emitted. Of course, no great accuracy can be maintained in locating the filament in the bulb, and it is for this reason that focusing means must be provided. It has been held that, so far as the projected beam is concerned, there is little advantage in using bulbs of more than 15 candle power with the sizes of reflectors permissible on automobiles, for the reason that with larger bulbs a great part of the filament is at such a distance from the focus that the rays from it when reflected add only to the roadside illumination directly in front of the car. This, however, may not apply to the new types of concentrated filament bulbs.

As regards the efficiency of the bulbs, it can be increased by increasing the voltage applied to the lamps. A slight increase in voltage over that for which the bulb is designed will result in a material increase in the light given out—a much greater proportional increase than the increase in energy consumption. But the life of the bulb is reduced thereby, and what is gained in current economy is lost in the wastage of lamps. When the bulbs are operated at their rated voltage a good balance is usually obtained between current economy and bulb life. The life of coil filament tungsten bulbs of 15 c. p. varies from 100 to 300 hours of burning. The following table, taken from Schroeder's paper mentioned above, gives the bulb characteristics with a fully charged and fully discharged battery.

Battery Condition	Per Cent Normal Volts	Per Cent Normal Amperes	Per Cent Normal Watts	Per Cent Normal C.P.	Per Cent Normal W.P.C.
Fully discharged (1.8 volts per cell)	68	79	54	23	233
Fully charged (2.6 volts per cell)	120	111	133	190	70

Mean Spherical Candle Power—It can readily be seen that owing to the different views obtained of the filament, according to the angle from which it is looked at, the light radiation varies with the direction. With a plain loop filament, for instance, the radiation of light is different in a direction perpendicular to the plane of the loop than in a direction parallel thereto. Therefore, if a single test for candle power were made, the result would depend upon the angular position of the filament relative to the photometer. The real criterion of the luminous power of a lamp is the mean of the candle powers shown in all possible directions, which is known as the mean spherical candle power. This is obtained by means of a special photometer. In order to obtain the mean intensity in a plane perpendicular to the axis of the bulb, the latter is rapidly rotated around the axis while in the photometer. To get the mean in a plane perpendicular to the former plane, the axis of the bulb is tilted against the axis of the photometer at different angles varying by steps of 10 degrees. A test is made for each position and the mean of all the readings is then taken. Observations need be taken only over a half circle, because the readings over the other half circle would give a diagram symmetrical to the first. A polar diagram of light distribution in a plane through the bulb axis is shown in Fig. 249 and it will be seen from this that the radiation is

LAMPS AND FITTINGS

strongest in a direction deviating about 15 degrees from the axis of the bulb opposite to the bulb base. A simpler method of determining candle power is to take measurements only in a plane perpendicular to the bulb axis, and the average of these readings is known as the mean horizontal candle power. Bulb manufacturers prefer to rate their bulbs by the current consumed rather than by the candle power, which is a purely nominal value.

Types of Lamps Used—It was formerly the custom to carry five outside lights on an automobile, viz., two headlights to illuminate the road ahead, two side lights to serve as signal lights outlining the width of the vehicle, and a tail light. The side lights also add to the illumination of the roadside close to the vehicle and take the place of the headlights when little forward illumination is required, as when the car stands by the curb. Since the introduction of electric lighting, separate side lamps have been discarded. Now side lamps are combined with the head lamps or none are carried. However, with the advent of electric lighting additional lamps have come into general use. Thus a dash lamp is now generally fitted to permit of reading at night the instruments mounted on the dashboard. Closed vehicles are provided with dome lights or special corner lights, and in some instances step lamps have been fitted under a translucent step, which are automatically lighted up when the door to the rear compartment is opened.

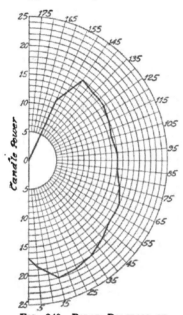

FIG. 249—POLAR DIAGRAM OF LIGHT DISTRIBUTION

Forms of Sockets—There are essentially two types of incandescent lamp bases and sockets therefor. The first is the screw socket, which is familiar to every one through its almost universal use in lamps for interior illumination. It is rarely used for automobile lighting. The second is the bayonet socket, which is commonly used for automobile lamps. The bayonet lamp base consists of a metal shell with two short oppositely located pins projecting therefrom radially. There are two corresponding slots in the shell of the socket, these extending lengthwise for about 3/8 inch, then circumferentially for a short distance and then lengthwise again in a reverse direction, thus forming a sort of hook.

Lamp sockets, connector plugs and bulb bases are also made in two types according to the system of wiring used, the

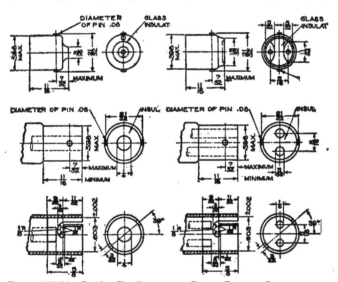

Figs. 250-51—S. A. E. Standard Bulb Bases, Sockets and Connector Plugs for Ground Return and Insulated Return Systems

ground return or single wire system and the insulated return or two-wire system. Bulbs, sockets, etc., used in connection with the former are known as single contact and those used in connection with the latter as double contact. Bulb bases, plug ends and sockets have been standardized by the Society of Automotive Engineers and the two types are shown in Figs. 250 and 251 respectively. There are two insulated contact points on the base of bulbs intended for the two wire system, but there is only a single central insulated contact point in the base of bulbs intended for the ground return system of wir-

LAMPS AND FITTINGS

ing. It will be understood that in the "two wire" lamp the ends of the filament are in electric connection with the two contact points, respectively, the filament being insulated from the metal portions of the base, while in the "ground return" lamp one end of the filament is connected to the central contact point and the other to the shell of the base.

Every lamp requires these three fittings—a socket secured in the lamp shell, a bulb base entering the socket at the inside of the lamp shell and a connector plug connecting to it at the outside.

Construction of Socket—The construction of the socket is clearly shown in Fig. 252. Inside the metal shell is fastened a plug of hard rubber. Two longitudinal holes are drilled through this plug and in each is clamped a hollow brass pin with a cap nut. In the head of the pin and in the nut the holes

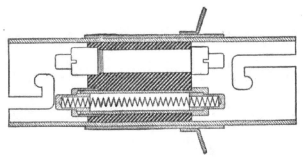

FIG. 252—SECTION THROUGH SOCKET

are somewhat reduced and through these reduced holes extend the shanks of two brass pins with a coiled steel wire spring between them pressing them outwardly. In forcing the bulb down into position the pins are pressed part way into the plug, and when the bulb is then rotated to bring it into the locked position the pins press against the metal contact plates on the bulb base.

The construction of a "ground return" socket is the same except that there is only one spring-pressed contact pin, which is located centrally in the socket. On returning, the current passes from the shell of the bulb base to the shell of the socket, which is grounded.

Connector Plugs—In making connection to a lamp socket, use is made of a connector plug, as illustrated in Fig. 253. This fastens into the bottom of the socket in the same way as the bulb fastens into the top thereof. It consists merely of a

308 LAMPS AND FITTINGS

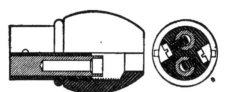

FIG. 253—SECTION THROUGH CONNECTOR PLUG

hard rubber plug into which are secured two short brass rods which are flush with the rubber at the top end but extend slightly beyond the rubber at the bottom end, where they are drilled and counterbored to receive the ends of the connecting wires. Brass set screws sunk into the rubber firmly hold the wires in place. A threaded hard rubber cap screws over the threaded lower end of the connector plug, the wires leading out through a central hole in this cap.

A single wire cable terminal, as used for Bosch head lamps, is shown in Fig. 254. After the insulation has been peeled off the end of the cable, a short length of brass tube is slipped over the end of the insulation and is fixed against lengthwise motion by forming two circumferential depressions in it, as shown. Then a gland nut is slipped over the cable and a cap is applied to the end of same, which latter consists of two brass parts insulated from each other, but clamped together by an inner thin brass tube expanded at both ends—viz., a ferrule, against the outer face of which the individual wires of the cable are applied, and a gland with two keys on it which engage in keyways in the fixture so as to keep the cap and cable from turning when the gland nut is screwed up. The lower portion of the gland is split and its end tapered to correspond to the tapering bottom of the counterbore in the gland nut, the arrangement evidently being intended to act as a lock for the gland nut.

Metallic Reflectors—Parabolic reflectors, if they are metallic, are generally made of drawn brass or copper, but sometimes of cast aluminum. The interior surface is heavily plated with nickel or silver and is then carefully burnished. The lamp front or door, containing the glass, must make an

FIG. 254—SINGLE WIRE TYPE OF CONNECTOR (BOSCH)

air-tight joint so as to exclude dust and moisture and prevent circulation of air. This is necessary in order to prevent tarnishing of the highly burnished reflector surface and resulting decrease in the efficiency of the reflector. In order to get the best illuminating effect the center of the filament must coincide with the focus of the reflector, and, since different bulbs may vary slightly in their dimensions, it is necessary to provide adjusting means for the bulb. In the Gray & Davis headlight these take the form shown in Fig. 255. The socket is secured in place by a clamp screw and when

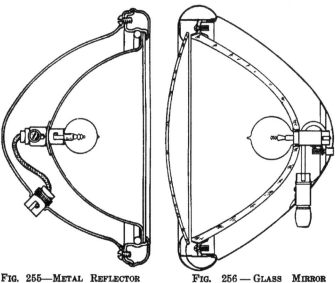

FIG. 255—METAL REFLECTOR LAMP   FIG. 256 — GLASS MIRROR REFLECTOR LAMP

this is loosened the lamp can be moved back and forth at will, after the lamp door has been removed. Adjustment can be properly made only by noting the effect on the road, and the best place to focus a lamp is on a dark road. The proper way is to focus the reflected light into a small ray or pencil beam, then properly align this ray with the road by resetting the lamp brackets, making sure that the car stands parallel with the road. Then adjust the bulb so as to give the degree of dispersion of light desired.

Glass Reflectors—Glass mirror reflectors are also used for automobile headlights. These are generally of the so-called "golden glow" type, which are made of a special glass having a greenish golden color. The light reflected by this lens is of a golden hue and is claimed to penetrate a foggy atmosphere to a much greater distance than a white or violet light. The back of the reflector is "silvered" in the same way as a mirror. The source of light, of course, sends out rays of all colors; but since the reflection takes place at the "silvered" surface at the back of the reflector, the reflected light must pass through the glass, and in doing so the violet and blue rays are absorbed, while the yellow rays are reflected.

Also, it is a well known fact that when light rays pass from one medium into another they are broken or refracted, as is most easily illustrated by holding a rod part way in water, when it will seem to be broken at the surface of the water. Since the light rays pass from the air into the glass and after reflection from the glass into the air, refraction plays a part in determining the direction of the reflected rays. For this reason the outside of the glass reflector usually has a section other than a parabola and is not parallel to the inner surface. One advantage of glass reflectors is that they never tarnish and never need to be polished.

In Fig. 256 is illustrated the Esterline head lamp, which has a "golden glow" type glass reflector. The reflector is clamped in two stamped metal rings which are held together by screws, with a felt packing between the rings and between the bearing part of the rear ring and the reflector. The front glass of the lamp is retained by tempered steel springs snapped into place and the complete front cover is held in place by two concealed latches held on by thumb screws. The cover makes continuous contact with the gasket above referred to, thus sealing the lamp.

Attention may be called to the focusing device, which is operated by a thumb wheel at the back of the lamp. This wheel operates a screw in a threaded lug fastened to the lamp socket, which latter is adapted to slide in openings in the center of the reflector and in a plate brazed into the back of the housing, respectively. An attachment plug is fastened into the bottom of the housing and connects by an inside cable to the socket.

Importance of Accurate Focusing—The subject of properly focusing the bulb in the reflector has been dealt with by G. L. Sealey of the Engineering Department of the National Lamp Works of the General Electric Co. Mr. Sealey recommends a method somewhat different from that outlined in the foregoing.

LAMPS AND FITTINGS

When automobile headlamps are properly focused, the center of the filament of the incandescent lamp coincides with the focal point of the reflector. When this condition is attained, the resultant beam is concentrated, dark spots are eliminated, objectionable stray light is diminished and the greatest amount of useful light is obtained. The sketch, Fig. 257, shows simply how the light is distributed by a parabolic reflector for three positions of a point light source. A', A', are parallel beams distributed when the point source is at the focus, while B', B', and C', C', are the beams dis-

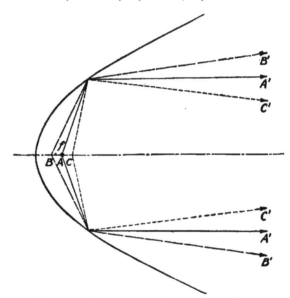

FIG. 257—REFLECTION OF OUT-OF-FOCUS RAYS

tributed when the point source is, respectively, back of and in front of the focus. Although the filament of an incandescent lamp is not exactly a point source, the concentrated filament of a mazda C lamp acts in the same general way as a theoretical point source. When the incandescent lamp is at the focus of a parabolic reflector, a beam of small diameter is directed forward which, when intercepted perpendicularly by a large card or paper, will give a light spot (Fig. 258). When the incandescent lamp is in front of or back of the focus, the light spot on the card will be as shown in Fig. 259.

312   LAMPS AND FITTINGS

An ideal location for focusing headlamps is a straight roadway having at least 75 ft. between the car and the side of a garage, or other surface vertical to the road, on which is drawn a line or mark 42 in. above the road level. When such a spot is found, the car should be placed so as to have the headlamps facing the side of the garage and projecting the light upon it. The relative position of the filament and focal point of the reflector should then be changed by moving the incandescent lamp (by means of the socket adjustment) forward or backward until the light spot of smallest diameter, or greatest concentration, is secured, as shown in Fig. 258. This operation should be performed on one headlamp at a time so as to be certain that each headlamp is correctly focused.

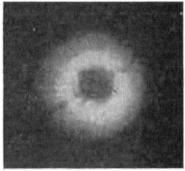

FIG. 258 — CONCENTRATED LIGHT SPOT INDICATING PROPER FOCUS

FIG. 259—CIRCLE OF LIGHT INDICATING IMPROPER FOCUS

Types of Focusing Devices—The four principal types of automobile headlamp focusing devices, or socket adjustments, are: (1) rim, (2) external, (3) bulb, (4) rear. In the rim type of adjustment the incandescent lamp is moved forward or backward by turning a screw placed on the rim of the reflector, and in the external type of adjustment this adjusting screw is placed to the rear of the lamp housing. In the bulb type, adjustment is accomplished by twisting or turning the incandescent lamp bulb. It is necessary to remove the reflector from the headlamp, and to loosen a screw or other means of holding the lamp in position at the rear of the reflector before adjustment can be made with the rear type. The latter two devices are not considered satisfactory since they necessitate the direct handling of the incandescent lamp or the dismantling of the headlamp in order to secure the desired result.

When the filament has been brought to the proper position with reference to the focal point of the reflector, that is, when the lamp has been properly focused, the height of the main beam of light above the level of the roadway should be carefully observed. Most state laws require that at 75 ft. in front of the car the main beam of light shall rise not more than 42 in. above the level of the roadway under any condition of loading. By carefully observing the line, or mark, on the side of the garage and the beam of light obtained from the headlamps this law may be complied with. It is often necessary to tilt the headlamps slightly by bending the supports in order to accomplish the desired result.

Care of Metal Reflectors—Reflectors may get dimmer with age, owing either to the reflector surface becoming dirty or becoming dulled or tarnished. Owing to the fact that the silvered surface is very easily scratched, it must not be wiped with a cloth. Any dirt should be washed out with a low-pressure jet of cold water and the reflector allowed to dry by evaporation in the atmosphere. If the silvered surface becomes tarnished it may be polished by means of crocus or jeweler's rouge, which is applied with a chamois cloth. The latter should be specially soft and should be so handled that no wrinkles can form in it. This is best accomplished by stretching the chamois over a wad of cotton waste. In polishing, the cloth should be moved in circles concentric with the axis of the reflector. Any other motion will make any scratches that may be formed on the polished surface stand out very conspicuously.

Mounting of Head Lamps—Head lamps, as a rule, are mounted on brackets secured to the frame side members in front of the radiator. The lamps should preferably be placed comparatively high, as otherwise any objects in the road or protuberances of the road surface throw exceedingly long shadows which give the driver the impression as if the roadway were full of holes. This makes the brackets quite long, and to prevent trouble from vibration a tie rod is often interposed between the two lamps. When no side lamps are used, and the head lamps also serve the purpose of indicating the width of the car to oncoming traffic, it is well to place them farther out to the sides than over the front spring horns, and a tendency in that direction has been noticeable in recent years. Thus the head-lamp brackets now are sometimes forged integral with the fender irons or secured to the sides of the radiator housing, while in a number of cars the lamps are built together with the front fender, which places them in what is perhaps the best position from the standpoint of oncoming traffic, and, besides, entirely eliminates lamp brackets. Generally the head lamps are inclined slightly downward, so that the shaft of light strikes the roadway about a hundred feet ahead of the car. At this point the beams have increased to such a diameter that the beams from the two lamps overlap, and there is no diminution in the intensity of illumination at the center of the path followed by the car. In fact, the intensity of illumination is likely to be greatest near the center and to decrease toward both sides, as shown by the diagram, Fig. 260. It will be seen that in this diagram the intensity of illumination is expressed in "apparent candlepower," a term which it may be well to explain. Photometric measurements were made at a distance of fifty feet from the head lamps. The intensities shown by the curve were obtained by applying the rule of the inverse squares of distances when the photometer showed equal intensity of illumination on the screen. But since the intensity of a projected beam does not vary strictly in inverse proportion to the square of the distance, the indicated candlepower is called "apparent." The "apparent candlepower" has a definite meaning only if the distance from the source at which observations were taken is stated. In Fig. 260 the dotted curves show the illumination from the single lamps and the full line curve the resultant illumination. It is well to so adjust the head lamps or their brackets that there is no appreciable diminution in the intensity of illumination at the center of the path where the beam strikes the roadway.

In a few cases, single headlights have been used, built into the radiator. This construction tends toward greater smoothness of outline of the car and toward current economy, but it necessitates the use of separate sidelights, as most state laws require "at least two lamps throwing a white light ahead." On cars and trucks used for commercial purposes in the city a single spot light is sometimes carried back of the windshield on an adjustable bracket, similar to searchlight brackets used on motor boats.

For the sake of neatness the connecting cables are sometimes carried into the lamp through the lamp bracket, so that they are invisible. Such concealed connections not only improve the appearance of the lamp and of the car as a whole, but facilitate cleaning of the lamps. Another improvement along the same line, tending to smoothen the contour of the lamp, is the combination of the supporting lug with the door hinge.

LAMPS AND FITTINGS 315

Tail Lights—Most of the state laws require that tail lights must throw a red light to the rear, and these lights therefore are generally provided with a red lens or glass, while some tail lamps consist of a spherical shell of ruby glass with a window of clear glass which throws a white light on the rear number plate. In connection with the tail light it is worthy of note that several state laws forbid arrangements whereby the light can be controlled from the driver's seat. Of

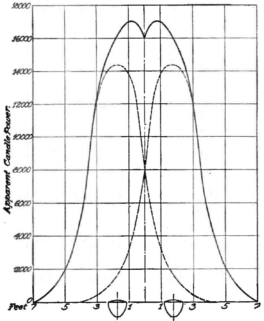

FIG. 260—HORIZONTAL DISTRIBUTION OF LIGHT FROM A PAIR OF HEADLAMPS AT A DISTANCE OF 50 FEET

course, it is permissible to have a switch convenient to the driver's seat whereby all of the lights can be turned on or off simultaneously, but the tail light must not be extinguishable separately by such a switch. A difficulty that has been experienced, mainly abroad, is that motorists have been summoned to court for driving at night with their tail light out when they believed it was burning. This may be caused either by a broken filament or some defect in the wiring. To guard against this eventuality some concerns connect a lamp on the dash in series with the tail lamp, using, for instance, two six-volt lamps on a twelve-volt circuit. Then, if the tail light fails for any reason, the other light also goes out and thus warns the driver. Some tail lights are made with two side glasses, one throwing a light on the number plate as required by law, and the other illuminating the roadside in such a manner that the operator can see whether it is burning. This arrangement fits in particularly with the now common practice of left drive.

Spot Lamps and Inspection Lamps—Spot lamps are small projectors which are supported on a bracket secured to the windshield or in some other position where they can be readily reached by the driver. They give an intense beam of light and

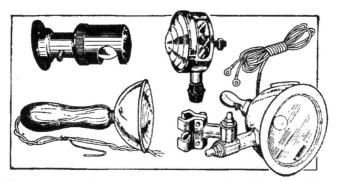

FIG. 261—DASH LAMP, INSPECTION LAMP, TAIL LAMP AND SPOT LAMP

can be turned in any direction, consequently they can be used to advantage for reading road signs, house numbers, etc., or lighting up any object not in the direct path of the car. Inspection lamps are small lamps provided with a metal guard and a length of cord or cable, which can be used for locating trouble anywhere on the car in the dark. As a rule, neither a spot light nor a trouble lamp is part of the regular equipment of a car, but some manufacturers make provision for their use in laying out their wiring systems, providing a convenient point of attachment. Recently a number of combined cowl and inspection lamps have been brought out.

Objectionable Glare—Even when acetylene head lamps were still commonly used on automobiles there was considerable objection to their blinding glare, and many drivers in the big cities then pasted translucent paper on the back of the lenses or glasses, or gave the lenses a coating of paint, except for a small central circle. When the still more powerful electric headlights became popular, so much annoyance was caused to pedestrians and drivers that several municipalities took action in the matter. One of the first cities to prohibit the use of glaring headlights in its streets was Chicago, whose ordinance provides in substance that "it shall be unlawful for any person operating an automobile to use a bright headlight, unless such headlight be properly shaded so as not to blind or dazzle other users of the highway." The New York City ordinance contains practically the same provision. The city of Cleveland has adopted an ordinance providing that at a distance of seventy-five feet or more ahead of the vehicle none of the reflected light from a headlight must be visible more than three feet above the roadway. A similar law is in force in the State of New Jersey.

This principle of limiting the maximum height of the shaft of light works all right on level ground, but it is ineffective when a car approaches the crest of a hill, which is one of the critical conditions in night driving. In such a case it would certainly be better if the driver had some means at his command for instantly reducing the intensity of the projected beam.

Cause of Glare—It would thus appear that the problem can be solved in two essentially different ways. Either the light must be dimmed as a whole or else it must be tilted or shaded in such a manner that none of its reflected rays can rise beyond a certain height. In this connection it may be well to explain what is meant by "glare," the term most frequently used to express the blinding effect of powerful headlights. Perhaps the best definition yet given is the following: "A glaring light is one which interferes with the acuteness of vision of adjacent objects." Glare is due chiefly to the ultra violet rays of the spectrum. It has therefore been proposed to use an amber colored lens or front glass on head lamps, which absorbs the ultra violet and blue rays and transmits only red, orange, yellow and green rays. This special glass used for the purpose transmits the red and other rays with very little absorption, hence the total radiation is not materially reduced. Another fact to be taken into account is that the red rays penetrate farthest through a misty or foggy atmosphere, as is shown by the fact that the sun when rising or setting always appears red. Hence the penetration of the beam of the headlight is not much, if any, diminished by the amber glass.

LAMPS AND FITTINGS

Dimming Devices—An early method of dimming electric headlights consisted in reducing the voltage applied to them, either by connecting the two lights in series across the battery or by introducing a resistance in the circuit. Connecting the lights in series is advantageous on account of the current economy resulting therefrom. There is one objection to it, however, namely, that in case one filament breaks both lamps will go out instantly and the driver therefore will be enveloped in more or less darkness. This, however, is a less serious matter in city driving than it would be in country driving, because of the street lighting.

By using two bulbs in the headlights, a larger one in focus and a smaller one out of focus, as shown in Fig. 262, both an intense light for country driving and a subdued, non-glaring light for city driving can be obtained from a single lamp

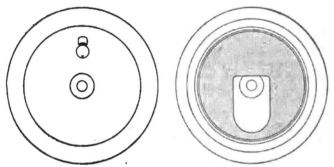

FIG. 262—DOUBLE BULB LAMP   FIG. 263—PARTLY FROSTED LENS

without any shading device and without waste of current. This arrangement gives the car the equipment of both head and side lights, reducing the side lamps to the small bulbs.

Control of Illumination—When the demand for glare control first became acute, various mechanical devices, involving shades and shutters of translucent material, were placed on the market, but these never came into extensive use, partly on account of mechanical weaknesses and partly because, while effectively eliminating glare, they cut down the road illumination too much. Another suggestion was to cover that portion of the bulb around the tip with a shield, so as to cut off the direct rays from the filament. This was evidently based on a misconception of the problems involved. It was the Lighting Division of the S. A. E. Standards Committee which first formulated the postulate that in order to obtain satisfactory road illumination on a level road no strong beams of light must be higher than 42 inches above the road surface at a distance of 75 feet or more ahead of the car. This led research and development into new channels. It was realized that the amount of light thrown upon the road must not be reduced any more than absolutely necessary and from this point on the endeavor was to restrict the field of intense lighting rather than to reduce the total radiation of light.

In connection with the directions for focusing head lamp bulbs it was brought out that beams from that part of the filament ahead of the focus, when reflected, approach the axis

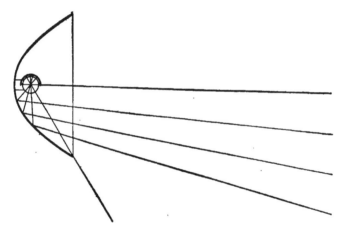

FIG. 264—HEMISPHERICAL REFLECTOR OR INTENSIFIER OVER BULB

of the reflector and may cross it. Thus, if the bulb is placed forward of the focus, the lower half of the reflector will throw the light upward, whereas the upper half will throw it downward. Similarly, if the bulb is placed back of the focus, the lower half of the reflector will throw the light downward and the upper half will throw it upward. This led to the development of methods for cutting off the reflection from one-half of the reflector and then regulating the direction of the light by adjusting the bulb so the filament was out of focus.

Intensifying Hemispherical Reflector—One method of cutting off light from one-half of the lamp reflector consists in placing a small hemispherical reflector either beneath or above the filament. If it is absolutely concentric with the center of the filament it will reflect any light falling upon it back upon the filament and thus intensify the source of light. All of the light emitted by the source, with the exception of the small amount absorbed by the hemispherical reflector, will therefore either fall upon the top half of the lamp reflector or pass directly through the front lens. Therefore, if the bulb is adjusted forward of the focus a very strong beam will be thrown on the road at a considerable distance ahead. The chief objection to such an under-mounted intensifier is that it cuts off direct radiation onto the road immediately ahead of the car. There is, of course, also, direct radiation from the bulb and reflected light from the hemispherical reflector in an upward direction through the lens, but this is rather too close to the car to be of harm.

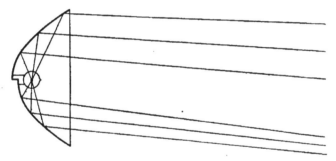

FIG. 265—HALVES OF REFLECTOR DISPLACED

By placing the hemispherical reflector over the bulb and placing the latter back of the focus these objections are overcome. The upper half of the lamp reflector does not receive any light, and the lower half, which receives the intensified radiation, throws it downward onto the road. Direct rays from the filament and the hemispherical reflector illuminate the road immediately in front of the car. Of course, the filament being out of focus, the light is somewhat dispersed and not as intense a beam is obtained as without controlling device and the bulb correctly focused.

Halves of Reflector Displaced—Another idea along the same line is to displace the top half of the parabolic reflector with respect to the bottom half. The top half may be so placed that the rear end of the filament is in its focus and the bottom

LAMPS AND FITTINGS

half so the forward end of the filament is in its focus. Then practically all of the filament is ahead of the focus of the top half, and the light will be reflected by this half in a downward direction. Similarly practically all of the filament will be to the rear of the focus of the lower half, and the light will be reflected from this half also downwardly. Hence the beams from the two halves of the reflector will intensify each other.

Prismatic Lenses—During 1914 and 1915 many special lenses for head lamps were placed on the market, partly in response to legislation in several states calling for the elimination of headlight glare. The Osgood (Fig. 266) is a plain lens with prismatic formations on the inside. These prisms refract the light downward onto the road. The extreme lower portion of the lens is plane, so as not to interfere with direct radiation from the bulb onto the road directly in front of the car, and a segment next to this plane one has a very slight prismatic formation.

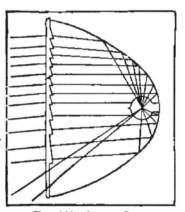

Fig. 266—Osgood Lens

The McKee lens, which is molded of amber-colored glass, has prismatic formations on the outside covering about two-thirds of the surface of the lens, while on the upper part there are pronounced up-and-down prisms which scatter the light falling upon this part of the lens, the object being to light up the sides of the road immediately ahead of the car and at the same time break up the upwardly directed parts of the projected beam.

Another plan which has been applied, but which seems to possess little merit, is to provide slats with reflecting surfaces of such width and so inclined that a small portion of the horizontal rays from the reflector pass through between the slats while virtually all the upwardly directed rays and most of the others are intercepted by the slats and turned downward or absorbed. It is fairly evident that this device makes it impossible to get strong road illumination at a considerable distance, as most of the rays deflected by the slats strike the road not far ahead of the car.

From the above it will be seen that there are two classes of devices intended to obviate annoying glare—the dimmers which are controlled from the driver's seat and those devices (mainly special lenses) which modify the beam of light given out and are permanent in their effect and beyond the control of the driver. Legislation favors these latter devices. All of these reduce the intensity of road illumination.

Illumination and Glare Tests—In order to determine the intensity of illumination necessary to reveal to an automobile driver on a dark night the presence in the road of a pedestrian in dark clothes at 150 ft. and 250 ft. respectively, and also the maximum glare which is considered tolerable by different persons, a series of tests were made in New York on the evening of March 5, 1918, under the auspices of the Illuminating Engineering Society and the Society of Automotive Engineers. About fifty observers served in these tests. In the results published the intensity of illumination is expressed in foot-candles, that is, the intensity of the source in apparent candle powers, divided by the square of the distance from the source in feet. The results arrived at by different observers varied enormously. The averages of all observations were as follows. To see a man in dark clothing at 150 ft. requires 8200 c.p. or 0.142 ft.-c.; to see a man in dark clothing at 250 ft. requires 6980 c.p. or 0.112 ft.-c.; the maximum glare permissible at a distance of 100 ft. corresponds to 239 c.p. or 0.024 ft.-c. The ranges of observation under the different headings were as follows: Visibility at 150 ft., 1000-10,000 c.p., 0.0445-0.445 ft.-c.; visibility at 250 ft., 1300-18,300 c.p., 0.021-0.293 ft.-c.; tolerable glare at 100 ft., 80-850 c.p., 0.008-0.085 ft.-c. These tests were made to form a basis for a New York State vehicle lighting law.

CHAPTER XXVI

Starting Motors

As has been stated previously, every direct current electric generator will act as an electric motor when current from an outside source is sent through its field and armature. While the principle of the machine as regards its generator action has already been explained, it will not be amiss to explain its action as a motor.

Principle of Action—Consider a two pole electro magnet whose poles are designated in Fig. 267 by N and S. Between these poles is located a coil which is carried on a shaft so as to be capable of rotation around its axis. When current is sent through the coil a magnetic north pole is formed, say, at N', and a south pole at S'. In accordance with the law of magnetic attraction the coil then will rotate in the direction indicated by the arrow until the magnetic axis of the coil coincides with the magnetic axis of the field. Then, obviously, there can be no further motion. But by using two coils at right angles to each other with their four leads brought out to a four section commutator, as shown in Fig. 268, we can get continuous rotation. In the position shown current flows through the coil lying in a horizontal plane, and since the magnetic axis of this coil is at right angles to the magnetic lines of the field, a torque will be exerted on the coil and it will begin to rotate. After it has rotated through 45 degrees from the position shown in the cut, the commutator brushes

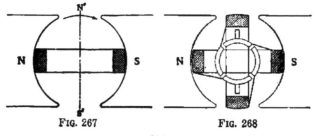

FIG. 267　　　　　　FIG. 268

pass from one set of segments to the other, and the other coil then becomes energized. In this way the magnetic axis of the armature is switched back 90 degrees and the rotation of the armature continues.

In Fig. 268 we have all the elements of a regular electric motor, only that the latter has many more coils and all of them are carrying current all the time. The resultant magnetic flux due to the armature current is always substantially at right angles to the field flux, and is switched back through an angle corresponding to one commutator segment every time the brushes pass from one segment to the next. There is, therefore, a continuous torque which is practically uniform.

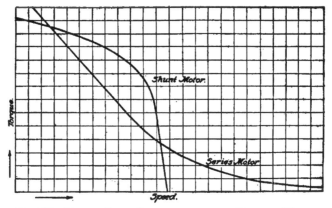

FIG. 269—TORQUE-SPEED CHARACTERISTICS OF ELECTRIC MOTORS

Types of Motors—The mechanical details of construction of an electric motor are absolutely the same as those of a generator. For instance, all of the different armature windings described in the chapter on Generators are also applicable to motors. Similarly, the different field windings used on generators are also used on motors, that is, there are series motors, shunt motors and compound or differentially wound motors. The relative adaptability of the different types of motors for engine starting is clearly indicated by their characteristic curves. In Fig. 269 are shown the characteristic curves of both a series and a shunt motor on constant potential circuits. Ordinates represent torques and abscissas speeds of revolution. It will be seen that in a shunt motor, as the torque increases there is a slight reduction in speed. Since the line voltage is assumed constant the field strength of the motor will be constant and the reduction in speed will be due mainly to voltage drop in the armature with increasing load.

In a series motor, on the other hand, the speed decreases rapidly with an increase in torque. In fact, a series motor, if not loaded, will race wildly and its armature may even burst from its excessive centrifugal force. This is due to the fact that when a series motor is running without load it takes very little current, and as the same current passes through the armature and fields, the fields are then very weak, hence the armature must run at an enormous speed to generate a counter-electromotive force approaching in value the applied electromotive force.

Starting Torque Important Factor—In the case of starter motors the most important factor is the starting torque or the torque exerted by the motor while at rest. In this respect the series motor is far superior to the shunt motor. When a motor connected to a battery is at rest it develops no counter-electromotive force and the battery, therefore, is virtually short-circuited. As in a series motor, the same current flows through armature and field coils, it follows that in such a motor when starting from rest an enormous current flows through the field coils and the magnetic field is saturated. This, together with the heavy armature current, insures a large starting torque. In a shunt motor, on the other hand, the field at starting is relatively weak, because the heavy current drawn by the armature results in a heavy drop in voltage in the battery and connection cables, consequently a lower voltage than normally will be applied to the field terminals and less current will flow through the field coils. It is particularly important that starting motors develop a strong starting torque, for the reason that it takes a greater torque to first start the engine in motion than to keep it in motion once it is started.

Compound Wound Motors—Where separate machines are used for current generation and starting, the starting motor is always series wound. When the same machine is used for both purposes it sometimes has two separate field windings, a shunt winding for use as generator and a series winding for use as motor. Some dynamotors are equipped with differential field windings in order to control the generator output. The series field winding when the machine acts as a generator then opposes the field shunt winding and weakens the field. On the other hand, when the machine acts as a motor, whereas the shunt current still flows in the same direction as when the machine acts as a generator, the series current now flows in the opposite direction so that the two magnetizing forces combine, instead of that due to the series winding partly neutralizing that due to the shunt winding. This is illustrated in Fig. 270, where the full line arrow shows the direction of generator current, while the dotted arrows show the direction of motor current. Such compound motors are used to a certain extent.

Counter-Electromotive Force—In an electric motor the torque developed is directly proportional to the number of lines of magnetic force passing through the armature, to the number of armature turns and to the armature current. Hence, the mechanical resistance which the motor must overcome—in other words, the torque it must develop—deter-

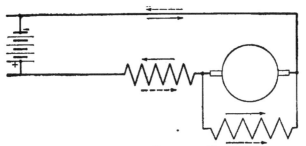

FIG. 270—CIRCUIT DIAGRAM OF COMPOUND-WOUND DYNAMOTOR

mines the current that will flow through the armature. This current, in overcoming the electrical resistance of the armature winding, causes a certain drop in voltage. The rest of the applied voltage is balanced by a counter-electromotive force generated in the motor armature. When a motor is carrying a normal load, the drop in voltage in the armature amounts to only a small percentage of the total, and the counter-electromotive force is then nearly equal to the applied electromotive force. As in the case of the induced electromotive force of a generator, the counter-electromotive force of a motor is proportional to the magnetic flux through the motor armature, the number of armature turns and the speed of revolution.

Power Required to Start Engines—The power required to start an engine depends not only on the bore and stroke and the number of cylinders, but also upon the compression carried, the clearances between bearing surfaces, the lubricants used and the temperature of the motor. There is no great variation in the torque required to start an engine from rest for temperature ranges above the freezing point, but there is a very marked increase when the atmospheric temperature drops down to zero. No general figures can be given regarding this increase in the starting torque, as it depends upon the character of lubricant employed and upon the accuracy of fitting the engine parts, but a good idea of its magnitude may be obtained from Fig. 271, which shows the variation of the starting torque with speed, of a four-cylinder 4¾ x

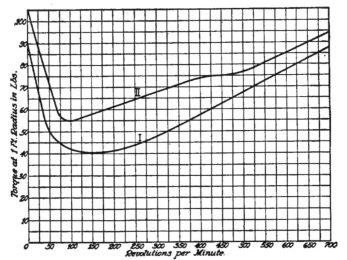

FIG. 271—ENGINE STARTING TORQUE AT NORMAL AND LOW TEMPERATURE

5¼-inch motor, with a compression ratio of 4:1—(I) at an atmospheric temperature of 75 degrees Fahr. and (II) when packed in ice. It will be seen from curve I that at normal temperature there is little variation in the cranking torque between 75 and 250 r.p.m., the torque being a minimum at 150 r.p.m. At the lower temperature the torque is a minimum at 100 r.p.m. The general form of all starting torque curves is the same, that is, the torque is a maximum when the engine is at rest and falls rapidly with a gain in speed until it reaches a minimum at from 50 to 150 r.p.m. Then the curve swings upward again, the torque increasing with the speed.

The average starting torque of an engine of a given total piston displacement is substantially the same whether the engine has four, six or eight cylinders, but the fluctuations in torque are, of course, greater the less the number of cylinders. Hence, the piston displacement being the same, a slightly smaller motor can be used with the engine of greater number of cylinders. A rough rule for starting torque is one pound-foot per 5 cubic inches total piston displacement, and for the cranking torque after the engine has once been set in motion one pound-foot per 10 cubic inches total piston displacement. It may be explained that one pound-foot is the torque represented by a force of one pound acting tangentially

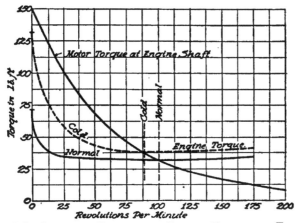

Fig. 272—Superposed Diagrams of Starter Torque and Engine Torque

at a radius of 1 foot. In practice, since the torque of which the starting motor is capable must exceed the torque required by the engine by a certain margin, the motor starting torque should be from 2½ to 3 times the normal cranking torque of the engine.

In Fig. 272 the torque curve of a starting motor multiplied by the gear reduction ratio is superposed upon the engine starting torque curves for normal temperature and low temperature, respectively. It will be seen that the starting torque of the motor exceeds the starting torque of the engine under cold weather conditions by a fair margin, as should be the case. The cranking speeds are shown by the points of intersection of the motor torque curve with the engine torque

STARTING MOTORS

curves, indicating an engine speed of 88 r.p.m. when the engine is cold and 100 r.p.m. when the engine is at normal temperature.

Cranking Speed—The main factor determining the most desirable cranking speed is the carburetion. Practically all carburetors have the characteristic of producing a lean mixture, that is, a mixture poor in gasoline, when the suction is weak, i. e., at low engine speed. Such a mixture ignites only with difficulty or not at all, according to the proportions. At a cranking speed of 100 r.p.m. or more a good mixture is generally formed, even at low temperature, when not all the fuel emitted by the carburetor nozzle vaporizes, and such a cranking speed is therefore desirable. If a magneto spark is used for firing the charge in starting (now a rather rare practice in this country) then the best cranking speed is also affected by the fact that a magneto gives sparks only above a certain minimum speed and the volume and igniting power of the sparks increase with the speed. However, battery ignition in starting is now almost universal and the battery spark has its maximum firing value when the speed is low. Another thing that requires consideration in this connection is that if the cranking speed is too low there is apt to be shattering due to back lash in the gear teeth. As the engine passes the compression point the compression naturally tends to accelerate it, and if this accelerating action is greater than the natural acceleration of the starting motor, then the teeth of the driven gear will move out of contact with the teeth of the driving pinion as the charge in the cylinder expands, and a shattering action results. According to one authority continuous driving action is obtained with four-cylinder motors above 80 r.p.m. and with six-cylinder motors above 50 r.p.m.

Energy Consumed in Starting—If an engine is in good working order so that an explosion occurs on one of the first compressions the energy consumed in starting is exceedingly small, as the following calculations will show. After two or three compressions the engine is up to its normal cranking speed and the current down to the normal value corresponding to that speed. Suppose we have a six-cylinder 3 x 5-inch motor, which has a piston displacement of 215 cubic inches. Hence, in accordance with the rules given above, the starting torque will be about

$$\frac{215}{5} = 43 \text{ lbs.-ft.}$$

and the cranking torque

$$\frac{215}{10} = 21.5 \text{ lbs.-ft.}$$

STARTING MOTORS

The average torque during the first three revolutions, therefore, is about 32 lbs.-ft. Since a point at a radius of 1 foot travels 6.28 feet in one revolution and 18.84 feet in three revolutions, the energy consumed during these three revolutions is

$$32 \times 18.84 = 602 \text{ foot-pounds.}$$

Since 2655 foot-pounds is equal to one watt-hour, converting the above mechanical energy units into electric energy units we have

$$\frac{602}{2655} = 0.226 \text{ watt-hour.}$$

This is the energy actually imparted to the engine. But the combined efficiency of the electric motor and the gearing would hardly be more than 70 per cent., hence the energy input into the motor would be

$$\frac{0.226}{0.70} = 0.32 \text{ watt-hour.}$$

If, therefore, we had a battery capacity of 480 watt-hours we could make

$$\frac{480}{0.32} = 1500 \text{ starts}$$

on one charge of the battery. The conditions here assumed are rather ideal. If the engine started every time on the sixth compression, instead of on the third, then the number of starts per charge would be reduced to 900.

It will be noticed that in the above calculations the cranking speed does not enter. In fact, the energy consumed is the same whether the cranking speed is high or low, provided the engine in each case picks up after a certain number of compressions. At high cranking speed more power is, of course, required for cranking, but the time of cranking is reduced in proportion, hence the energy, which is the product of power and time, remains the same.

In Fig. 273 is shown an oscillogram of the current consumed by a motor in starting an engine. This was taken by Prof. Benj. F. Bailey of the University of Michigan. The engine used in the test had four 3¾ x 4-inch cylinders (177 cu. ins. displacement) and was rated at 25 h.p. A Disco single unit machine with compound winding was used, and was connected to the engine with a gear ratio of 3:1, both when acting as a starting motor and when acting as a generator.

Referring to the curve, it will be seen that the first rush of current was 236 amperes, which is equal to 69 per cent. of the current that would have flown had the armature been held from rotation. The peak of the current curve indicates the time when the starter pulled the engine over the first compression. The time elapsing between the closing of the switch and the passing over the compression was 0.043 second. The torque exerted by the starter with a current of 236 amperes was 28 lb.-ft. or 84 lb.-ft. at the engine shaft. Tests showed that the torque required to pull the engine over the first compression was 35 lb.-ft. The difference, 39 lb.-ft., was used in overcoming the inertia of the engine and starter.

As soon as the engine was under way the current dropped to an average value of 63 amperes. The peaks represent the increase in current necessary to overcome the compression. The starter, of course, was assisted by the inertia of the flywheel. The time between peaks is 0.133 second, corresponding to an engine speed of 225 r.p.m. With a starter cranking the engine at a lower speed the peaks would have been more marked.

The ignition switch was not closed when the starter switch

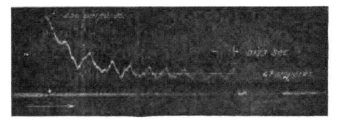

FIG. 273—OSCILLOGRAM OF STARTER CURRENT

was operated. At the point where the current drops suddenly to zero the ignition was thrown on and the engine began to fire. The time represented by the complete cycle of operations shown was 2.12 seconds. A normal start requires about half a second.

Starter Speed—There is an extreme range in the speeds of revolution of different types of electric starter motors, depending upon the method of drive employed. Some of these motors are direct connected to the engines to be started, the starter armature forming the flywheel of the engine. Since in this case starter speed and cranking speed are necessarily the same, the latter is chosen somewhat higher than with the other types of starters. Thus, in the U. S. L. system (Fig. 275), which is of this type, the cranking speed varies between 200 and 400 r.p.m. in the different cars to which it has been applied. Sometimes a single gear reduction to the flywheel is used, in which case a maximum reduction ratio of about 10:1 can be obtained and the motor speed may be anything between 800 and 1200 r.p.m.

A higher gear reduction is obtained with a double reduction set of spur gears to the flywheel, viz., from 20 to 30, and with such a gear the motor speed may be as high as 3000 r.p.m. In the past one maker of electric starters has been using a high reduction worm gearing with a ratio of about 70:1 and employed a motor running at something like 7000 r.p.m. while cranking the engine. Within certain limits the weight of a motor of a given output varies in inverse ratio to the speed. That is, a low speed motor is comparatively heavy and a high speed motor light in proportion to its output. The great weight of the very low speed direct connected motor is mitigated by the fact that its armature replaces the flywheel and it also eliminates the reducing gear. Further, since the drive is direct, with absolutely no bearings in the motor or between motor and engine, there is no loss in transmitting the starting torque. Of the two most directly competing systems, the single and double reduction gear drives, the former has the advantage of a simple and more efficient drive, the latter that of permitting the use of a lighter, more compact and probably somewhat more efficient motor.

Characteristic Curves of Motors—In Fig. 269 was shown a torque-speed curve of a series motor, as well as a similar curve of a shunt motor. In giving the characteristics of a starting motor it is usual, however, to plot on the same diagram curves showing the variation of the voltage at the motor terminals, of the speed of revolution, of the torque and of the efficiency, with the current in amperes flowing into the motor. Such a diagram of motor characteristics, applying to a Rushmore Model B starting motor which weighs 47 lbs. and when geared through a single reduction of 10:1 turns a six-cylinder 4½ x 5½-inch engine over at 108 r.p.m., is shown in Fig. 274. There are really two sets of characteristics shown in this diagram, one referring to a motor with 31 armature sections and the other to a motor with 27 armature sections, the different windings being used in order to obtain different speeds. Naturally the torque is greater with the greater number of sections, and the efficiency is also greater in that case. It will be seen that the torque increases in almost direct proportion to the armature current in amperes. The maximum efficiency with the low speed winding is attained at about 150 r.p.m. and amounts to about 72 per cent., while the maximum efficiency for the high speed winding is attained at about 175 r.p.m. and amounts to about 69 per cent. The horsepower developed may also be plotted on the same diagram.

Mechanical Construction—Owing to the exceedingly severe service to which a starting motor is subjected it should be very substantially built. The shaft must be of liberal

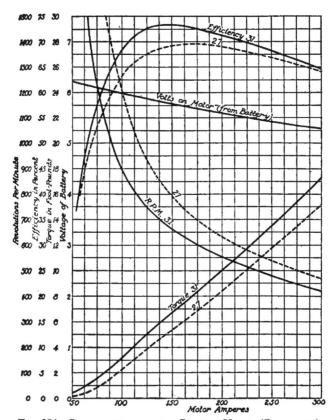

FIG. 274—CHARACTERISTICS OF A STARTER MOTOR (RUSHMORE)

diameter and the ball bearings in which it is supported should be of ample capacity. The armatures of starting motors are generally of the form wound type, since, with the low voltage employed, only a single pair of conductors is required per slot and the conductors are exceptionally heavy. Owing to the sudden current impulses when the starter switch is closed the motor should have a relatively strong field, so as to insure good inherent commutating characteristics. In that case low resistance brushes can be used without danger of destructive sparking at the commutator under the conditions of heavy overload obtaining while starting. The commutator also should be of liberal dimensions.

We are now in position to discuss the considerations which generally determine the question whether a single machine shall be used both for generating the electric energy and for starting the motor or a separate machine for each function. The advantages of a single machine are obvious. Ordinarily it weighs less than two separate machines and it requires only a single driving connection to the engine. The machine must be either shunt or differentially wound, in accordance with the requirements of battery charging. However, for starting purposes, a series machine is by far the best. Some makers have solved this difficulty by providing the machine with both a shunt and a series field winding, using one when the machine operates as a generator and the other when it operates as a motor. This, of course, considerably adds to the weight and bulk of the machine, as not only is the field winding duplicated, but the magnet poles must be made sufficiently long to accommodate the two windings.

Aside from the field winding, the armature winding and commutator also would not be made exactly the same for the generator as for the motor. For instance, in a 6-volt system, the generator delivers a current of about 15 amperes. As a motor, however, it will take a current at least ten times as great, and hence, even though this current flow is maintained only for an instant, the motor requires a larger commutator. Another thing is that the armature current moves the point of commutation around in a generator in the direction of armature motion—in a motor in the opposite direction. Hence, in a separate generator the brushes would be given an advance corresponding to the full output of current, while in a motor the brushes would be set back through a certain angle. It is evident that in a dynamotor a compromise has to be made and the brushes set substantially in the no load neutral position.

Classification of Single - Unit Systems — Single - unit machines or combined generators and starters may be divided into four classes, viz., (1) direct connected or flywheel type machines; (2) single-speed machines, gear or chain connected to the crankshaft at the same ratio whether operating as generator or motor; (3) machines having two different driving connections to the engine, from opposite ends, one generator connection which makes the dynamotor run slightly faster than the engine and one starter connection which makes the engine turn over a great deal slower than the dynamotor, and (4) machines embodying an automatic change gear and connecting to the engine through a chain or gear drive, the change gear being cut out of action when the drive is from the engine and into action when the drive is the other way. We will consider these types in succession.

CHAPTER XXVII

Starter Drives

In the following treatment of starter drives the various types are considered in the chronological order of their appearance. It should be pointed out, however, that the screw drive, invented by Vincent Bendix, has gained for itself a very important place and is used far more extensively than any of the others. In 1918 the manufacturers stated that 150 different car models had this starter drive.

Flywheel Type—This type of dynamotor or single-unit system is represented by the U. S. L. and Otho systems, the former of which will be briefly described here (see Figs. 275 and 276). On account of the fact that the dynamotor must develop a torque greater than the starting torque of the engine it will naturally be quite bulky and heavy, but since its revolving member takes the place of the flywheel, this objection partly vanishes. In order to get as much flywheel effect as possible from a given weight the armature is placed outside a star-shaped magnet frame having eight poles. The armature, therefore, is of ring form and has a so-called ring winding, being wound with insulated copper ribbon, which lies in armature slots on the inside of the ring and on top of the outside of the ring. A 16-armed spider bolted to the rear end of the clutch drum, which in turn is bolted to the crankshaft flywheel flange, supports the armature as well as the commutator. The field frame is supported by an aluminum housing, which is bolted to the top half of the engine crank case. The brush supporting ring is clamped between the housing of the dynamotor and an annular cover plate, oblong slots being cut in the ring where the bolts pass through so as to permit angular adjustment of the ring. Some of the advantages of the direct driven, flywheel type of dynamotor are as follows:

Since the armature acts as a flywheel it represents no extra weight. No driving connection is required, hence the installation of the dynamotor is very simple, and noisy operation is obviated. The machine does not require any space at the side of the engine, hence does not interfere with engine accessibility.

STARTER DRIVES

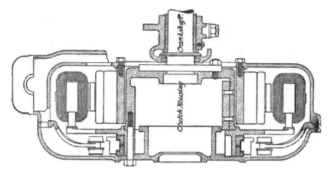

Fig. 275—U. S. L. Flywheel Type Dynamotor (Section)

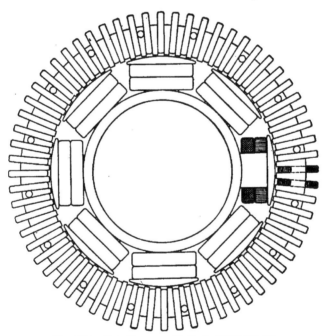

Fig. 276—End View of U. S. L. Dynamotor

338 STARTER DRIVES

The electrical efficiency of a flywheel type dynamotor is undoubtedly rather low, on account of the low speed and notwithstanding the fact that there are no bearing losses. This construction is now obsolete. The high cost of construction was against it.

Single-Ratio Drive—Coming next to the second class of single-unit systems, in which the dynamotor is geared to the crankshaft in the same ratio, whether acting as generator or motor, we have the Entz (Fig. 277) as the most prominent example. Owing to the fact that the machine must not run too fast when acting as generator, it cannot be geared up very much. A ratio of 3 to 1 is about the limit, and for high-speed motors the ratio might well be less. As a motor the machine must therefore develop a torque about one-third the engine starting torque, which, while much less than that re-

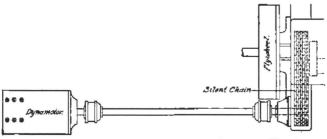

FIG. 277—SINGLE DRIVE DYNAMOTOR (CHALMERS-ENTZ)

quired of the flywheel type dynamotor, is much more than that required of separate starting motors which are geared down from 10 to 1 to as much as 30 to 1. Consequently a relatively large machine is required—much larger than necessary to carry the normal generator load of about 100 watts. The construction, however, is very simple, since there is merely a plain chain drive to the crankshaft, without the so-called overrunning clutch used on most starting motors, and the dynamotor, owing to its higher speed as compared with the flywheel type, can be made with fewer poles and considerably smaller and cheaper. While originally the single-drive ratio system was fitted mainly to cars in the higher-priced class, it has recently been developed by other manufacturers for use on light, low-priced cars, for which service its simplicity seems to make it particularly suited.

Two-Ratio Drive System—In the two systems so far described the required size of the electric machine is determined solely by the starting torque it must develop, as it requires a very much larger machine to develop on its shaft a torque equal to, say, one-third the starting torque of the average automobile engine than to produce at three times engine speed an output of, say, 100 watts as generator. In view of the great difference in the output required as a generator and as a motor if the gear ratio between engine and dynamotor is the same in both cases, the thought suggested itself of employing variable gearing between dynamotor and engine so that the dynamotor might run at high speed while cranking the engine and yet not at too high a speed when acting as generator. One method of accomplishing this purpose is employed

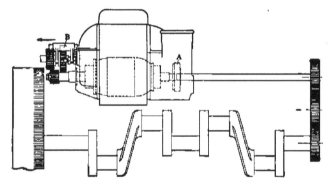

FIG. 278—DOUBLE DRIVE DYNAMOTOR (DELCO)

in the Delco system, in which there are two separate driving connections—one to the forward end of the crankshaft and the other through the flywheel to the rear end of the crankshaft. Two so-called overrunning clutches A and B (Fig. 278) are used with this system. These are in the nature of ratchets and permit of transmitting motion through them in one direction, but not in the other. Such clutches are well known through their use in lawn mowers, for instance. When the mower is pushed the revolving wheels impart motion to the cutter wheel; when it is pulled the wheels revolve idly. In Fig. 278 the forward overrunning clutch A is the "generator clutch," that is, the power is transmitted through it when the dynamotor is being driven from the engine. Looked at from the forward end of the car the crankshaft of the average engine revolves right-handedly and a shaft driven from it through a train of three spur gears, such as the generator drive shaft, also revolves right-handedly. Clutch A is so arranged that it will transmit motion from its central member to its outside member when the former revolves right-handedly. The armature of the dynamotor always rotates right-handedly, whether the machine acts as a generator or motor. Right-hand rotation, however, can be transmitted from the dynamotor only through clutch B. Therefore, when the dynamotor acts as a motor, it is disconnected from the forward end of the crankshaft by clutch A and it drives through clutch B, a set of sliding gears and the flywheel to the rear end of the

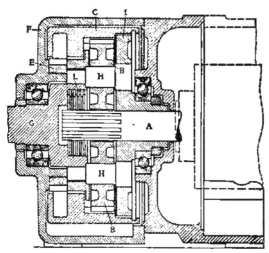

FIG. 279—JESCO AUTOMATIC CHANGE GEAR

crankshaft. As soon as the engine picks up its cycle it rotates faster in proportion than the armature of the electric machine, but it cannot drive the armature through the flywheel because clutch B will not transmit right-hand rotation from the rear toward the front. Instead, the motion of the crankshaft is now transmitted to the armature through clutch A. The forward drive may be such that the dynamotor as generator will make one and one-half revolutions to one of the crankshaft and the rear drive such that the dynamotor as motor will make 20 or more revolutions to one revolution of the crankshaft. Hence, the torque of the motor in starting is multiplied twenty

STARTER DRIVES 341

or more times by the gearing, and a small dynamotor can be used.

Automatic Change Gear—A variation of the type of double drive just described incorporates an automatic change gear in the dynamotor and employs only a single driving connection between dynamotor and engine crankshaft. The gear is operated or controlled by means of a centrifugal governor mounted on the driving shaft, which runs at low speed when the engine is being started and at high speed when the engine runs normally. A typical variable gear of this type, the Jesco, is illustrated by the two views in Figs. 279 and 280.

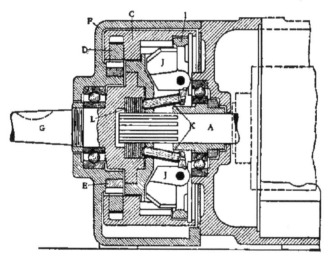

FIG. 280—ANOTHER SECTION OF THE JESCO GEAR

Referring to the drawings, A is the armature shaft, upon the end of which gear teeth are cut. The small pinion thus formed meshes with two planetary pinions BB, which in turn mesh with an internal gear C, forming part of what is known as the control casing. This casing carries a pair of pawls DD, adapted to engage with a ratchet wheel E, which is fast upon an inwardly projecting hub of housing F. The pawls are held in engagement with the ratchet wheels by flat springs.

Both the armature shaft A and the sprocket shaft G always rotate right-handedly. When current is admitted to the armature of the dynamotor for starting and the armature begins to turn, planetary pinions *BB* roll on internal gear *C*, and their bearing studs *HH*, being secured into a flange on sprocket shaft *G*, carry that shaft along in the same direction in which shaft *A* is rotating, but at a considerably reduced speed. Bearing studs *HH* are supported at their opposite ends in a disc *I*, which is supported by an annular bearing at the end of the motor housing, and which in turn serves as a bearing for armature shaft *A*. Disc *I* forms a support for two governor weights *JJ* in the form of bell cranks, the inwardly extending arms of these governor weights pressing against short push rods *KK*, which transmit the pressure to a miniature multiple disc clutch *L*.

Since disc *I* is rigidly connected to sprocket shaft *G* by means of the planetary bearing studs *HH*, the governor weights will be carried around at the same speed as sprocket shaft *G*, and when the speed attains a certain value the weights will fly out and thereby engage clutch *L*. It is apparent that armature shaft *A* is then direct connected to sprocket shaft *G*, and since by this time the engine has picked up its cycle and drives the dynamotor, the drive will be transmitted directly from shaft *G* to shaft *A*. The control casing now rotates in unison with the two shafts, the pawls slipping backward over the ratchet wheel.

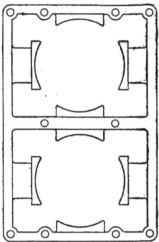

FIG. 281—DOUBLE DECK DYNAMOTOR FRAME

Recently there has been a decline in the use of systems employing a dynamotor with a variable gear connection to the crankshaft, which is probably due to a patent decision to the effect that this arrangement is covered by a basic patent issued to Coleman and which is controlled by the Dayton Electrical Laboratories Co.

Tandem and Double Deck Machines—The first step away from the strict single-unit system is that employing two machines combined in one. For instance, we may have the two field frames arranged one on top of the other in a single casting, as in Fig. 281. The generator armature is then geared to the crankshaft in the desired ratio, while the motor shaft is connected to the generator shaft through a train of gears in one of which is incorporated an overrunning clutch so arranged that power may be transmitted from the motor shaft to the generator shaft, but not the other way. The gearing is such that the motor shaft makes several turns to one of the generator shaft. If current is sent into the motor it will drive

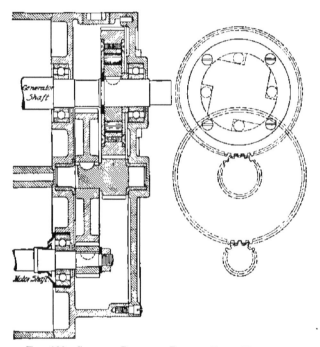

FIG. 282—STARTER DRIVE OF DOUBLE DECK MACHINE

through the generator and its drive. As soon as the engine picks up the generator will run ahead of the motor and when the current is shut off from the latter it will stop. A form of gear train employed with double deck machines is illustrated in Fig. 282.

Overruning Clutches—The most popular type of overrunning clutch is the roller type illustrated in Fig. 283. It comprises a central steel disk, in the circumference of which are cut three or four notches containing hardened steel rollers, and a steel drum surrounding the disk. When the rollers are in the position shown in full lines they are wedged in between the disk and drum, and, provided the angle between the tangents to the roller circle at the two contact points is sufficiently acute, any left-hand turning effort applied to the disk will be transmitted to the drum. In order to insure instant engagement of the rollers with the drum and disk surfaces when a turning effort is applied, small coiled springs are lodged in drill holes in the disk so as to lightly press the rollers into contact with the engaging surfaces of both disk and drum. If a left-hand torque is applied to the drum instead of the disk, friction between drum and rollers at once moves the roller to the released position.

Another type of overrunning clutch, as used on the Aplco system, is illustrated in Fig. 284. It is claimed that this design of clutch is more rugged than the roller type. Referring to the illustrations, A is the fan pulley keyed to the forward end of the crankshaft, with which the clutch is sometimes combined for simplicity's sake. Rigidly attached to this part are pins B which carry pawls C. Resting on pawls C are light springs D designed to prevent vibration of the pawls. The clutch also comprises a ratchet element made up of three parts riveted together and the whole is bolted to the sprocket wheel E, a casing F being used to enclose the pawl and ratchet elements to keep out dirt, etc. G represents the end of the gear case on the engine, which carries a hub that serves as a bearing for the driving member. The pawls and ratchets are so arranged that when the motor is idle there will be one of the pawls not more than 0.089 in. from engagement, with an 8-pawl element; consequently there is no shock at the moment of engagement or at the time of going over compression. When current is applied to the starting motor the driving member carrying the ratchets, one of which is in engagement with one of the pawls, begins to turn the pawl-carrying element, which is keyed to the crankshaft. When the engine picks up its own power the pawls are thrown back into the recesses provided and out of engagement with the ratchet wheel, and will stay there until the car is stopped, when they

FIG. 283 — ROLLER TYPE OVERRUNNING CLUTCH

are returned to position by springs D. However, to overcome any possibility of the pawls dropping down and making a noise when driving at a very low speed on direct drive, the noise-killing element H is introduced for the purpose of receiving the pawls as they drop down into position. The mounting of pawls C on pins B is such that no strain is put on pins B, the strain being carried by the abutment back of the pawl. It is seen that the driving member is idle while the engine is running.

Need of Shifting Gears With Flywheel Drives—In all the starting and lighting systems in which the starting effort is

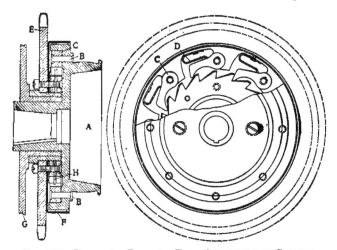

Fig. 284—Pawl and Ratchet Type Overrunning Clutch

transmitted to the crankshaft by the same device which transmits the power required for driving the generator there is no need for a mechanical disconnecting device between starting motor and crankshaft. This includes the system in which a dynamotor is permanently geared to the crankshaft, that in which a dynamotor with automatic change gear connects to the crankshaft by gear or chain, and that in which the motor and generator are arranged tandem-fashion, the motor driving through the generator. But whenever there is a separate starter drive from the motor to the flywheel, a pair of gears must be enmeshed previous to starting and unmeshed after starting. If the flywheel pinion remained in mesh with the

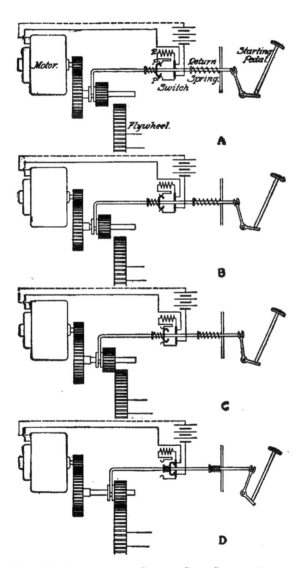

Fig. 285—Operation of Sliding Gear Starter Drive

flywheel gear it would be very noisy at high engine speeds, though the motor, of course, might be disconnected from the engine by an overrunning clutch. The enmeshing must be done at the same time that the starting switch is closed, and provision is always made to insure automatic meshing of the gears simultaneously with or immediately after closing the switch. There are three well-known mechanical methods of accomplishing this. The first consists in the provision of a pair of sliding gears of which one is adapted to engage with a flywheel gear and the other is in engagement with the starter pinion. In this way a double reduction is obtained between starter shaft and engine shaft.

Starter Resistance—In most systems employing sliding pinions the switch operates in two steps, closing the motor circuit first with a resistance in circuit and then with this resistance cut out. With the resistance in circuit the motor will turn over slowly without load, so that the gears can be meshed without difficulty. If the motor were connected at once directly across the battery terminals it would quickly attain a very high speed and then the gears would be hard to enmesh.

Operation of Shifting Gear and Switch—The successive positions of the shifting gear and of the starter switch are clearly shown in Fig. 285, which represents a Westinghouse system. At A is shown the "off" position of the shift pinion and switch contactor. Pressure on the starting lever moves the shift rod first to the position shown in B, closing the motor circuit at P and P^1 through the resistance R; this starts the motor at low speed. Further motion of the shift rod to position C opens the electric circuit, but the motor and pinion continue to turn, owing to their momentum. When position C is reached the pinion is still turning slowly, so that it cannot fail to mesh with the gear; but, as power is turned off the motor there is no difficulty in sliding the teeth into full engagement. As soon as the teeth do engage, further foot pressure on the starting lever shifts the rod to the position shown in D, closing the electric circuit at Q after the pinion and gear have meshed a sufficient distance to present a good bearing length on the teeth; this connects the motor directly to the storage battery, so that full power is impressed, and it turns the engine over until the starting lever is released or the engine picks up on its own power. There is an overrunning clutch between the flywheel pinion and the motor, so that if the pedal is not promptly released when the engine picks up, the motor is not driven by the engine. When the pressure is removed from the starting lever, the shifting-rod springs return all parts to position A; this releases the gears and opens the electric circuit, and the motor comes to rest. The travel of the switch rod, starting from clear "open" position, is approximately as follows: 3/16-inch, auxiliary contacts closed; ½-inch, auxiliary contacts open; 9/16-inch, gears start to mesh; 1¼-inch, main contacts close; 1⅜-inch, main contacts compressed.

In some cases the switch is not mounted directly on the gear-shift rod but is on a separate rod which acts on the gear-shift rod by some form of lever. In such cases the switch-shift rod should be provided with a separate spring, to insure that the switch is always returned to and held in the full-off position when the starting pedal is released.

Inertia Controlled Meshing Gear—In another design of starter drive, the Bendix (Fig. 286), use is made of the inertia of a weight on the starter shaft to force the pinion

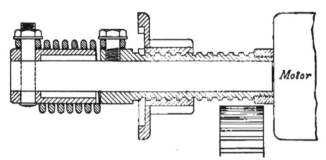

Fig. 286—Section Through Bendix Spring Drive

along the shaft when the latter begins to turn. The drive consists of a shaft having screw threads on the outside, and a gear pinion having corresponding threads on the inside, so that the gear screws over the shaft like a nut over a bolt. A circular weight, which is slightly out of balance, is fastened to the gear. A coiled spring connects the motor shaft to the hollow screw shaft.

When the electric motor starts, it drives through the spring and turns the screw shaft. Owing to the inertia of the weight secured to the gear, it cannot immediately follow the screw shaft, and since the gear does not turn it is forced to move along the shaft. This causes it to engage with the flywheel gear, and after that it keeps on moving along until it reaches the stop at the end of the screw shaft. The two gears are then fully meshed, and it is obvious that when the screw

STARTER DRIVES 349

gear has reached the stop it cannot move any farther along the shaft and must revolve with the screw shaft. At this point the screw shaft and electric motor are revolving at great speed and therefore have gained considerable momentum, which adds to the torque actually produced by the motor in starting the engine. The entire starting torque is taken through the coiled spring, which keeps coiling until all of the power has been applied to the flywheel gear.

As soon as the engine starts exploding and runs under its own power, the flywheel, of course, turns much faster than when it was cranked by the starter. This causes it to turn the screw gear at a higher rate than the revolution of the screw shaft. The result is that the gear will be screwed on

FIG. 287—BENDIX SPRING DRIVE, OUTSIDE VIEW

the threads of the shaft until it is out of mesh with the flywheel gear. This unmeshing movement is entirely automatic and eliminates the use of an overrunning clutch.

It might be supposed that as soon as the screw gear was out of mesh, if the electric motor kept on running, the former would be automatically screwed back into mesh with the flywheel gear. It is to obviate this that the weight on the screw gear is made unbalanced. The unbalanced weight twists or cocks the screw gear so that it binds on the screw shaft and turns with it. This automatic clutching is due to the centrifugal force on the unbalanced weight. When the electric motor stops running the screw gear has been fully moved away from the flywheel gear and it remains in that position until it is again required to start the engine.

One of the advantages claimed for the Bendix drive is that owing to the fact that the starting motor is running at a very high speed at the moment the engine first commences to rotate, a greater starting torque than that of which the motor is capable may be applied to "break" the engine loose.

Screw shift direct drives are made in two types, outboard and inboard. In the former the pinion when meshing with the flywheel gear travels away from the starter; in the inboard type the pinion travels toward the starter in meshing and with this type no outer bearing is required beyond the pinion.

Bijur Automatic Shift for Double Reduction Drives— Joseph Bijur has designed a number of modifications of the screw drive in which he has sought particularly to overcome the possibility of jamming of the pinion against the flywheel teeth. One of these designs (Fig. 288) is for a double reduction drive and the other (Fig. 289) for a single reduction drive. Fig. 288 shows a section through the housing W to which the starter is bolted, and which in turn is bolted into the flywheel housing by means of a flange. The starter shaft has long teeth X cut on its end, which project into the housing W. These teeth are always in mesh with those of driving gear D mounted on drive shaft B.

Pressing on the face of driving gear D is a leather or Thermoid clutch member E which is riveted to steel plate F. A light shell S extends around the clutch and serves to deflect dirt, oil, etc. The whole clutch assembly can move horizontally on the drive shaft but must rotate with it on account of flats C on the shaft which register with similar flats in clutch plate F and clutch face E. Adjacent to clutch plate F is a sleeve H with two slots through which a pin G extends, this pin serving to limit the horizontal travel of the clutch assembly to $5/8$ inch. A spring cover R holds pin G in place.

Spring L exerts pressure between the face of the drive gear and the clutch member E, and this pressure is sufficient to cause the gear to turn the clutch and shaft with it unless the shaft is held against turning. The pressure is not sufficient, however, to enable the clutch to transmit any real power.

In starting, by closing a single contact switch the starter is connected directly to the battery and it begins to rotate. Driving gear D rotates drive shaft B through the clutch mechanism. Pinion M screws itself to the left, and if the pinion teeth register with the spaces between the flywheel teeth the pinion will continue to travel into mesh until its entering face comes into contact with sleeve H. As the pinion

STARTER DRIVES

continues to travel towards the left, it pushes sleeve H, the whole clutch assembly and drive gear D in the same direction and compresses spring L. When shoulder Z of driving gear D comes in contact with shoulder V on the shaft the pinion is in full mesh with the flywheel.

With spring L tightly compressed, there is sufficient pressure between the face of drive gear D and clutch member E to cause the clutch to transmit to the drive shaft the power required for cranking. The starter then spins the engine until it takes up its own cycle. When the engine begins to fire, pinion M is rotated faster by the flywheel than by the starter. The pinion therefore travels to the right and out of mesh

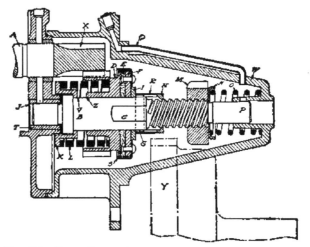

Fig. 288—Bijur Screw Shift for Double Reduction Drives

with the flywheel gear. On coming out of mesh it is cushioned by spring O.

If the pinion teeth and flywheel teeth should butt on being brought together, the action is as follows: On closing the starting switch pinion M travels to the left until its teeth butt with those of the flywheel. The pinion then is prevented from rotating, and, since the drive shaft is rotating, it screws itself through the pinion, moving to the right and carrying with it the sleeve H, clutch assembly, driving gear D and spring L. The drive shaft and parts assembled on it move to the right until sleeve H engages pinion M, and the pressure exerted between the pinion and sleeve then firmly clamps the two together. The pinion teeth are pressed against the flywheel teeth by the comparatively light spring O; therefore, when sleeve H firmly grips pinion M it causes the pinion to rotate with it. After a small rotary movement of the pinion its teeth register properly with the spaces between flywheel teeth, and the spring O snaps the drive shaft and parts assembled thereon to the original position occupied before any movement to the right occurred. The pinion is then free to travel into full mesh with the flywheel and start the engine.

Bijur Direct Geared Automatic Shift—Another screw shift has been designed by Mr. Bijur for single reduction drives. Referring to Fig. 289, pinion A is driven by notches on the inward flange on housing B registering with the pinion teeth. This housing has six slots about ¾ inch deep cut in its other end into which the notched clutch plate C registers. The cork clutch D is held in place in C by means of a notched retainer, and its right hand face can slip slightly against the smooth ground face of the flanged nut E whenever the torque exceeds the friction between the cork and steel faces, this friction being due to pressure exerted by spring F. Power is thus transmitted in the working position from the shaft to the flanged nut E, thence to clutch D, its retainer, clutch plate C, housing B and thus to pinion A.

In addition to the cork clutch there is a fixed clutch G mounted on the other end of the pinion flange. It is held against the flange by means of the clutch spring washer H, sleeve J, and wire lock spring K. The torque necessary to cause this clutch to slip is fixed irrespective of the position of the flanged nut.

Normally meshing is effected by depressing the starting switch, thereby causing the shaft to rotate within the flanged nut E, with the result that the nut travels along the threaded shaft and forces the entire assembly of the housing and pinion into the position of mesh through the spring F. This spring rests at its left-hand end against a flanged plate riveted to the pinion. When the nut gets to the end of its motion it first compresses a pair of dished washers through a pair of thrust bearing washers. These dished washers, together with clutch D, take up the shock of starting. When clutch plate C strikes the thrust bearing washer the axial pressure on the clutch face rises enormously and the clutch thereby becomes self-tightening.

When the pinion is thrown out of mesh by the excess speed imparted to it by the flywheel, the axial motion of the nut, clutch and housing is cushioned by a coiled spring over the nut.

STARTER DRIVES

Whenever a pinion tooth strikes the corresponding end of a flywheel tooth, nut E compresses spring F and urges the housing forward along the teeth of the pinion. Finally clutch plate C strikes the thrust washer, which tightens clutch D and causes nut E to turn with the shaft. Housing B is therefore turned by the notched clutch plate C and in turn transmits its rotative effort to the pinion, releasing the "butting" teeth. The instant this happens, spring F, which has been greatly compressed by the previous action, snaps the pinion into mesh with the flywheel, and normal cranking takes place.

The function of the weak spring S is to prevent "ticking in" of the housing and pinion especially when the car is running down hill.

Safety Devices—There is some danger of a driver inadvertently pressing on the starter button while the engine is

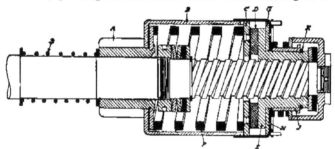

Fig. 289—Bijur Screw Shift for Direct Geared Drives

running. This causes severe clashing of the gear teeth, and, if frequently repeated, would quickly destroy the meshing gears. Some rather elaborate mechanisms have been devised to make gear meshing under such conditions impossible. A rather neat and simple arrangement for locking the shifting gear against accidental operation has been used by the Overland Company. On the housing of the shifter mechanism is located a solenoid actuated by current from the storage battery. The core of the solenoid forms a latch and when in its normal position prevents the shifting pinion from being enmeshed with the flywheel gear. The supply of current to the solenoid is controlled by a push button convenient to the driver's seat, and when the driver wishes to start the motor he first depresses the push button, which unlocks the shifting mechanism, and then depresses a heel button which effects the actual shifting.

STARTER DRIVES

Magnetic Meshing Gear—A unique method of meshing the starter gears is used in the Rushmore system of the Bosch Magneto Co. It employs a single reduction by spur gears to a gear ring on the flywheel. The armature of the starter motor, whose shaft carries the spur pinion which can be brought into mesh with the flywheel gear, is so arranged that it has a free longitudinal motion in its bearings equal to the motion required to unmesh the pinion. A sectional view of the motor is shown in Fig. 290. At the commutator end the armature shaft is drilled out to a depth of 10 inches, and a coil spring is inserted which bears with a pressure of about 15 lbs. against the hardened pin running against the cap of the bearing. The spring normally holds the armature in the position shown, when the pinion is out of mesh with the flywheel gear and the armature is partly out of the armature

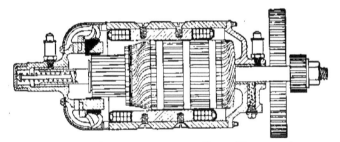

Fig. 290—Magnetic Meshing Gear (Rushmore)

tunnel. When the switch is closed the normal attraction of the field magnet poles draws the armature like the core of a solenoid into its working position, with the pinion properly engaged. The armature then starts to turn.

With this form of starter gear a special switch is used which causes the motor to turn over slowly when the circuit is first closed, thus facilitating meshing of the gears. Two views of the starter switch are shown in Fig. 291. In the position of the switch shown, the armature is short circuited. When the switch contact arm touches the middle contact a current of about 150 amperes flows from the battery through the motor field and the resistance strip, and while the switch arm remains on the first contact about half of the full current is shunted around the armature. Thus the current through the field instantly pulls the armature into the working position and the gears in mesh, while the momentary short circuit

STARTER DRIVES

across the armature prevents its rotating at high speed, although in the intense field the armature current of approximately 75 amperes produces a very heavy torque. Thus the armature rotates very slowly, and it is claimed that the gears can never fail to mesh.

When the switch arm leaves the first contact the armature short circuit is removed, but the resistance coil holds the battery current down to a value which, though sufficient to hold the armature in the working position, does not permit the motor to give its full torque. When the arm reaches the bottom contact the resistance is cut out and the motor exerts its full torque in turning over the engine until the explosions occur.

The bent-over spring at the left of the switch is the auto-

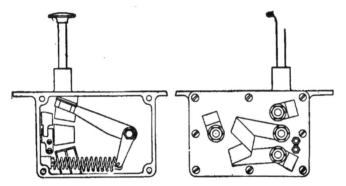

FIG. 291—SWITCH FOR MAGNETIC MESHING GEAR CONTROL

matic "ride-over." This spring rides over the flexible switch arm and tends to hold it against the middle contact during the downward movement. On the return movement, however, the flexible switch arm rides up on the ride-over spring or clip until it has passed the middle contact, then drops down to the original short circuit position only after opening the battery circuit. The result is that the battery circuit is opened before the armature is again short circuited. Otherwise, on the return movement of the switch the armature would draw the starter pinion into engagement with the rapidly moving flywheel. This would cause noise and would evidently be bad for the gears.

The instant the engine picks up, the motor is relieved of its load and the current drops to almost nothing. Then the solenoid action ceases and the spring instantly pushes the armature into its idle or free running position. Thereafter the current required to spin the armature is too small to draw the latter back into its working position, hence the armature and pinion spin idly until the operator releases the foot switch.

Starter Switches—Several types of starter switches have been illustrated already. They may generally be classed under one of two heads, viz., those giving only a single contact and those first closing the circuit through a resistance and then cutting out this resistance. One more switch of the single point type, the Auto-Lite, is shown in Fig. 292. It consists of a pressed steel box secured by wood screws to the under side of the toe-board. In the bottom of this box are secured two terminal bolts by means of nuts, insulating washers and bush-

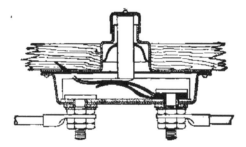

FIG. 292—SINGLE POINT STARTER SWITCH (AUTO-LITE)

ings. Under the head of one of these bolts is secured the switch arm or blade—a curved piece of strip brass which is urged upward by a flat steel spring below it. On the cover of the switch box there is a cylindrical projection which extends through a hole in the toe board and forms a guide for the switch plunger. A pressed steel cap is secured to the top end of the plunger and passes over the outside of the plunger bearing, thus preventing dust from getting into the bearing and into the switch box.

Another type of two-point switch, the Ward Leonard, is shown in Fig. 293. This is called the harpoon type, on account of the shape of the movable switch member. The cables connect to the terminal blocks in the upper corners of the switch box and these are directly connected to the two main switch blades *AB*, made of copper laminations. Below these there are two auxiliary switch blades *CD*. Blade *C*

STARTER DRIVES

is directly connected to blade A, but between blades B and D the starting resistance is inserted. Movable switch member E first establishes connection between the auxiliary switch blades, and the starting current must then flow through the starting resistance. As the movable member E is forced down farther, contact is established between the main switch blades, and the starting resistance is then cut out of circuit.

Brush Switch—In the Delco system no separate starting switch is used, but instead, one of the motor commutator brushes is ordinarily raised out of contact with the commutator and is let down upon it when it is desired to start the engine. This same brush also acts as a generator switch, closing the generator circuit when it is up. It will be remembered that this is a single-unit system, the same machine serving both as generator and motor, but it has two armature windings, two commutators and two sets of brushes.

Referring to Fig. 294, when the starting lever is pulled backward, rod A is pushed forward, causing gear B of the starting clutch to mesh with motor pinion C. Immediately after gears B and C are in mesh, gear D, which is integral with B, meshes with the gear teeth on the flywheel and at the same time the extension of rod A to the bell crank E allows the motor brush F to move toward the commutator, breaking the generator circuit at G and closing the motor circuit. When the starting lever is released a spring throws the gears out of mesh and at the same time raises the brush from the commutator and closes the generator circuit.

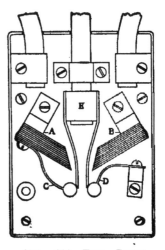

FIG. 293—Two Point Starter Switch (Harpoon)

There are two common locations for the starter switch, viz., under the footboard directly underneath the driver's foot, or on top of the starting motor. With the former arrangement the driver presses directly on the switch rod and no operating linkage is required; with the latter the length of the electric cable in the starter circuit is materially shortened and thus the drop in voltage in the cables at starting is minimized.

Starter Troubles—If, when the starter switch is closed, the starter fails to revolve, it may be due to any one of several causes, as follows: The battery may be almost discharged or entirely dead; there may be poor contact at the battery, switch or motor terminals; the starting switch may not make good contact with the switch blades; the motor brushes may be stuck in the brushholders and out of contact with the commutator. Any of these disarrangements would be likely to prevent the motor from starting, irrespective of the drive employed. Of course, a "frozen" engine is another possible cause of failure to start. This may be due either to extremely low

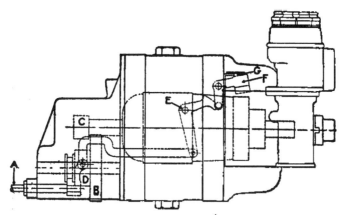

FIG. 294—DELCO BRUSH SWITCH

atmospheric temperature and the use of heavy cylinder oil, or to the motor having been greatly overheated during the last previous run. In the former case heating of the engine, as by filling the cooling system with boiling water, will overcome the trouble; in the latter case the piston can generally be loosened by pouring some kerosene into the cylinders and allowing it time to act.

With a "frozen" engine the action of the starter is somewhat different, according to the method of drive. If the drive is through a train of gears remaining permanently in mesh, the starting motor will not turn at all. On the other hand, if there is a flywheel drive so that the pinion has to be enmeshed before the engine can begin to turn, the motor will rotate for a moment before it becomes connected with the engine and if it can be heard revolving this indicates that the trouble is with the engine, rather than with the starter, although it does not prove this, as a weak battery might give sufficient current to turn over the motor idly, but not enough to enable the motor to start the engine. In all such cases it is therefore best to get out the starting crank and crank the engine by hand to see whether it turns over with only the normal resistance or whether it is frozen.

Symptoms of Trouble—If the starter turns when the switch is closed, but the engine is not set in motion, and if a test with the hand crank shows the engine to turn over with the normal freedom, it may be due to the overrunning clutch slipping. This will sometimes occur if the clutch is improperly oiled or if clutch members, such as pawls or rollers, have broken.

With systems employing flywheel drive, if the gears sometimes mesh readily and at other times fail to mesh, the reason may be that the auxiliary switch contacts which close the circuit through the starting resistance do not make good contact, that the starting resistance is burned out or has loose or dirty connections. The result of any of these derangements would be that no current would flow through the motor with the switch in the intermediate position, hence the motor would not turn over at low speed so the gears could be meshed easily.

It will sometimes occur that a starting motor with automatic gear shift will continue to run after the engine has picked up. This is generally due to the fact that the switch return spring is not strong enough.

With an automatic gear shift, if the starter rotates when the switch is closed but the engine does not it is a sign that the pinion is not drawn into mesh, which may be due to too tight a fit on its shaft or to marring of the gear teeth preventing proper meshing. Where a screw shift is employed a moderate amount of lubricant must be put on the starter shaft.

In no case should the starter be run for any considerable time if the engine fails to pick-up, because this is a heavy strain on the battery. If the motor fails to operate there is something wrong with it and the cause of the trouble should be found and eliminated before the battery is run down.

CHAPTER XXVIII

Fuses, Lighting Switches, Wiring Fixtures and Wiring

Fuses—There is always a possibility of a short circuit occurring in the car wiring, and as this would lead to an enormous current flow in the wires, the latter would quickly become heated and possibly fused. To obviate this a "weak link" is inserted in each circuit in the form of a fuse. The working element of a fuse is a short length of lead wire of a diameter depending upon the maximum current to be carried. If this current is materially exceeded, the

FIG. 295—ENCLOSED TYPE OF FUSE

lead wire fuses and the circuit is broken. In order to prevent the fused lead to be thrown about and possibly causing an explosion, and also in order to permit of quick replacement of "blown" fuses, the fuse wire is now enclosed in a glass tube with metallic ferrules at the ends. These are known as enclosed type fuses (Fig. 295). The transparent housing enables one to see at once whether the fuse is good or whether its fuse wire has been melted.

Fuses are now made of standard capacities, for maximum currents of 10, 20 and 30 amperes. There are two standard sizes of fuses, viz., 1¼ inch in length over all, with ferrules ¼ inch in diameter, and 1½ inches in length over all, with ferrules 13/32 inch in diameter. As regards the capacity rating of fuses, the National Board of Fire Underwriters specifies that with the surrounding atmosphere at a temperature of 75 degrees Fahrenheit they must carry indefinitely a current 10 per cent. greater than that at which they are rated, and at a current 25 per cent. greater than the rating they must open the circuit without reaching a temperature which will injure the fuse, tube or terminals of the fuse block. With a current 50 per cent. greater than the rating and at a room

FUSES, LIGHTING SWITCHES, ETC.

temperature of 75 degrees Fahrenheit, the fuses, starting cold, must blow within one minute.

Fuses are held in metal fuse clips as illustrated in Fig. 296, which depicts a fuse block of Westinghouse manufacture. The clips must be so designed that the fuses cannot slip out accidentally and must be fastened to their base in such a manner that they cannot turn. Further, they must be so designed that they cannot be sprung together or apart so much that the material will take a permanent set. If bent together it would interfere with the insertion of the fuse, and if bent apart the clip would not hold the fuse firmly.

All circuits carrying current from the storage battery, with

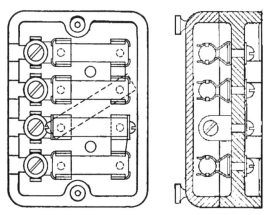

FIG. 296—FUSE BLOCK

the exception of the starter circuit, should be protected by fuses. There is no need for fuses in the starter circuit because this circuit is closed only for very short periods; besides, the maximum value of the current that ought to flow in this circuit and the time the current should flow are so indefinite that it would be difficult to specify the capacity of a fuse which would give real protection to the battery, circuit, etc., and not give constant trouble from blowing when there is nothing wrong with the wiring.

The fuses must be inserted on the battery side of all switches and junction boxes. The reason for this is that the closer they are to the battery the greater the protection they give. For instance, if the fuse blocks were located beyond the switches and junction boxes, any short circuit in these devices would not cause the fuses to blow and the battery would not be protected. In laying out the circuits the fuses must be so located that the blowing of a single one will not extinguish all of the driving lights. It is obvious that if all lights were extinguished simultaneously on a dark road while the car was being driven at a good clip, considerable danger would be involved. It is also desirable that no matter what fuse blows there is always a live connection plug to which an inspection lamp cord can be attached.

Lighting Switches—On a great many cars individual switches are used for each lamp circuit—one for the head lamps, one for the dimmer and one for the dash and tail lamps. The switches most generally used are either of the "push" type or of the "push-and-pull" type. Both of these switches are made in gangs of from 2 to 4.

The mechanism of a push switch is illustrated in Fig. 297. From the center of the base A of insulating material rises a pressed steel pillar B to which is pivoted at about midheight the pressing C in the form of a three-armed lever. One laterally extending arm of this lever carries the switch blade D which is insulated from it. The latter is adapted to make contact between two sets of contact blades secured into blocks carried by the insulated base and carrying terminal screws at the bottom of the base. The downwardly extending arm has its end bent at right angles and a coiled spring E hooked to it. The other end of this spring is anchored to the top of the pillar B, whose end is bent over for the purpose. To the lateral arms of the triple armed lever are pivoted plungers FF which extend through holes in the switch cover. It will readily be seen that in the position shown the spring holds the switch firmly in the closed position; but if the plunger which is now up is pressed down, as soon as the triple lever passes its central position, the spring pulls the other way, and thereafter it holds the switch firmly open. This "snap" feature is embodied in all lighting switches. It prevents the switch being left in a position where arcing would be likely to occur, that is with the blade and contacts in close proximity. This switch, as well as all others for the same purpose, has only two positions of equilibrium, viz., fully open and fully closed.

Fig. 298 shows a sectional view of a "push and pull" type of switch, the Connecticut. It consists of a hollow cylindrical block of molded insulating composition. Into the bottom of the block are molded two brass blocks into which

FUSES, LIGHTING SWITCHES, ETC.

two hard brass clips are soldered. A plug of insulating material with a brass tip extends through a hole in the brass cover plate, having a button head at its outer end. The brass tip of the plunger is rounded off and there are two circular grooves on the plunger, one in the brass and one in the composition. When the plunger is pressed in, the "necked" portion of the clips bears on the insulating material of the plunger and the switch is open; when the plunger is pulled

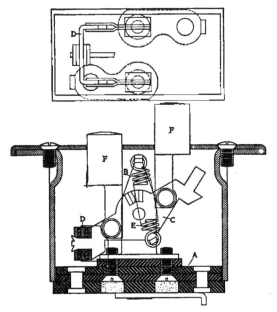

FIG. 297—MECHANISM OF PUSH SWITCH (H & H)

out, the necked portion of the clips bears on the brass tip of the plunger and the switch is closed. In each case the necked portion of the clips rests in one of the grooves of the plunger, which latter is held firmly in each position and passes with a decided snap from one to the other. Brass screw sockets for the binding screws and the screws holding the brass cover plate in place are molded into the composition.

In the Gray & Davis switch, which is of the combination type, the fuses are placed on the rear of the switch, thus

FUSES, LIGHTING SWITCHES, ETC.

doing away with a separate fuse block. Fig. 299 is a rear view of the switch, showing all the electrical connections. It will be seen that there are eight metal plates on the back of the switch which are divided into four pairs, each pair being connected together by a fuse. The letters on the back of the switch have the following meanings: B—battery; H—headlights; S—sidelights; R—rear light. From the four plates at the top, bottom and two sides, posts extend up into the switch box where they carry switch contacts. The spindle of the switch carries a number of switch blades that are insulated from it and from each other, which make connec-

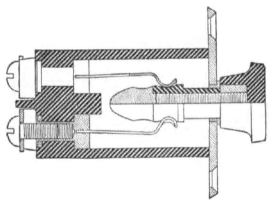

FIG. 298—SECTIONAL VIEW OF PUSH AND PULL SWITCH (CONNECTICUT)

tion between different posts in the different positions of the switch handle.

Fig. 300 shows a gang switch as used with Delco systems for controlling both the lights and ignition. The individual switches are of the push-and-pull type and they are mounted on a plate which fastens to the instrument board or dashboard. The two switches on the right are for the ignition, connecting the circuit either to the storage battery or to an emergency battery of dry cells, respectively. The other three switches control the lighting circuits—one the tail light and cowl light circuit, another the head light circuit, and the third the dimmer resistance, which is incorporated in the switch. A lock is provided for the ignition switches. This consists merely of a metal block which by means of the key

FUSES, LIGHTING SWITCHES, ETC. 365

can be slid back and forth behind the switch plate. In one position of the locking block the switches can be pulled outward through holes in the block, while in the other position of the block this is impossible, the stem of the switch then

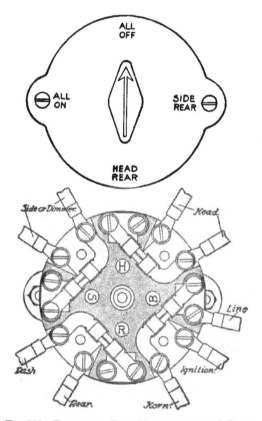

FIG. 299—FRONT AND REAR VIEWS OF GRAY & DAVIS LIGHTING SWITCH

being in a slot of the block so narrow that the contact ring cannot pass through.

Ground Return and Insulated Return Wiring Systems— There are two general methods of wiring, known respectively as the ground return and the insulated return, or the single wire and the double wire system. With the ground return system one terminal of the generator and the battery, as well as of each current consuming device, is grounded, while insulated wires connect the other terminal of each consuming device through a switch, fuse and possibly an ammeter to the current source. Hence the current, after flowing from the battery through the insulated wire to the consuming device, passes from it into the metal frame work of the car and returns through that to the battery.

In the insulated return system, on the other hand, the current flows from the battery to the consuming devices through one wire and returns through another wire. One advantage of the ground return system that is immediately apparent is that with it only one-half as much wire is required. The chief advantage of the insulated return system is that it affords better protection against short circuits. With a ground return system, if any portion of the insulated wire or any fitting becomes grounded, the battery is at once short circuited. On the other hand, with an insulated return system a ground on one side of the circuit does not cause any trouble. It is only when both sides of the circuit are grounded that a short circuit is produced. With a ground return system, if the gap between any exposed current carrying fitting and a part of the frame is accidentally bridged by a screw driver or other tool a short circuit is caused. One important advantage of the ground return system is that all fittings designed for it, such as bulbs, bases and connector plugs, require only a single insulated contact and therefore can be made much more substantial. It is not possible to use a dash lamp and tail lamp in series on a ground return system. Or, perhaps it is more correct to say, if the dash lamp and tail lamp are to be connected in series, one of these lamps at least must be of a type designed for insulated return systems. Of course all the rest of the lamps can be wired on the ground return system.

It is recommended by the Society of Automotive Engineers that the storage battery be grounded on the positive side and by only one conductor. This ground should be made in an accessible place by not more than a single screw or nut or else it should be made readily accessible by means of an accessible switch. The motor circuit usually has an insulated return, but if a ground return should be used the ground connections must be of a very substantial character on account of the heavy currents to be carried. A good method of making the motor ground connection consists in riveting and

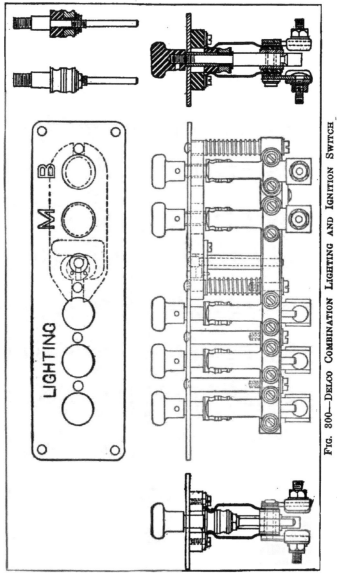

FIG. 300—Delco Combination Lighting and Ignition Switch

FUSES, LIGHTING SWITCHES, ETC.

soldering a brass or copper plate to the frame, after the latter has first been thoroughly scraped. A terminal lug is sweated to the end of the cable and is then securely bolted to the plate.

Wiring Systems—In Fig. 301 is shown a diagram of a complete installation wired on the ground return system. This is a 12-volt system and the head lights and side lights, both using 12-volt bulbs, are connected in parallel respectively. Owing to the advantage of having the tail light and dash light in series, these are provided with 6-volt bulbs and connected in series. This necessitates that one or the other of these two lamps be a two contact lamp as ordinarily used with insulated return systems. There are, therefore, really three lighting circuits and each of these is controlled by a switch and protected by a fuse. The circuit from the generator leads

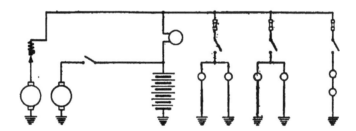

FIG. 301—GROUND RETURN WIRING SYSTEM.

through the automatic circuit breaker, and the starter circuit is completed through the starter switch. No fuse is required in the starter circuit, for reasons already pointed out. An ammeter, if one is fitted, is inserted in the battery circuit just beyond the starter connection, and the ammeter will indicate both charging and discharging current except current passing into the starter. By connecting the ammeter between the battery terminal and the junction of the starter cable it could be made to indicate starting currents also, but these currents are so large as compared with the ordinary charge and discharge current that if the instruments were made with such a scale as to indicate these latter with a fair degree of precision, it would be likely to be injured by the heavy starting currents, and if it were made with a scale of sufficient range to comprise the starting current it would not show the

FUSES, LIGHTING SWITCHES, ETC.

ordinary charge and discharge currents with any degree of accuracy.

A wiring system embodying the insulated return principle is shown in Fig. 302. The only differences between the two systems are that all fittings are of the type having two insulated contacts, two wires are run to each fitting, instead of one, and no part of the system is grounded.

Three Wire System—In Fig. 303 is shown what is known as the three wire system. A 12-volt battery is used, but all lamps are of the 6-volt type and are connected in series across the mains. In addition to the two main wires there is a third wire leading from the middle of the battery and connecting to the junction of each pair of lamps in series, except the tail and dash lamps. The advantage of the three

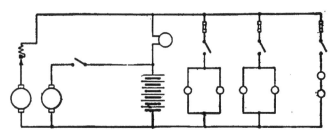

FIG. 302—INSULATED RETURN WIRING SYSTEM

wire system may be explained as follows: A 12-volt battery is preferred, because if the same size of wire is used the energy loss in the cables is only one-fourth as great, or if the system is designed for the same energy loss much smaller wires can be used. But 12-volt lamps are not so reliable as 6-volt lamps of the same candle power, because the filaments of the former must be longer and thinner.

By connecting two lamps in series, 6-volt lamps can be used on a 12-volt battery, but if a filament in one of the bulbs fails, both lamps become extinguished simultaneously, which involves an element of danger. This is obviated by the three wire system. With this system there is ordinarily no current flowing in the central wire, but if one filament breaks, or a fuse blows, the lamp in series with the one disabled receives its current through the third wire and continues to burn.

370 FUSES, LIGHTING SWITCHES, ETC.

Inasmuch as the cells of the storage battery are then being discharged unequally it is not advisable to continue running in this way for any length of time, and the broken bulb or blown fuse should be replaced at the earliest opportunity. It will be seen that with this wiring system a fuse is needed for each lamp, instead of for every pair of lamps, as in the two previously described systems, and a two-pole switch instead of a single pole switch must be used for each pair of lamps.

Fuses and Switches Required—In Fig. 304 is shown a system, which, while employing an insulated return, permits of the use of single contact lamp sockets and fittings. The two lamps of each pair are connected with their insulated terminals to opposite terminals of the battery and the other terminals of both lamps are grounded. This system, too, requires a fuse for each lamp and a double pole switch for each pair of lamps. If one bulb breaks or a fuse blows, the

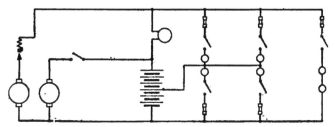

FIG. 303—THREE WIRE WIRING SYSTEM.

system becomes unbalanced, and if it happens to be a headlight that is thus incapacitated the chances are that the sidelight on that side of the system will be burned out as it will be subjected to a greatly increased voltage, while the voltage on the remaining two lamps will be reduced and they will burn dimly. While in this system one side of the fittings is grounded, an accidental ground on one side of the circuit will not involve the fire hazards that it would in a ground return system, for whereas in the latter system such a ground would short circuit the battery, in the former it would not. However, a ground on one side of the line would throw the whole pressure of 12 volts across one set of 6-volt lamps and the bulbs of these lamps would burn out. Of course, the ground return system is partly protected against dangerous arcing in case of a short circuit by means of the fuses, and these should preferably be close to the battery, as the fuses give

FUSES, LIGHTING SWITCHES, ETC. 371

no protection against short circuits on that part of the line between them and the battery. In some cases a special main line fuse is used in addition to the fuses in the separate circuits and inserted into the line close to the battery.

Wiring diagrams shown in Figs. 102 and 103 are equally adaptable to 6-volt systems, by using 6-volt bulbs for head and side lamps and 3-volt bulbs for the tail and dash lamps or else using 6-volt bulbs for these lamps also and connecting them in parallel.

S. A. E. Installation Specifications—Following are the specifications for ground return electrical installations issued by the Society of Automotive Engineers:

Use of Conduit.—Insulated conductors must be protected by metallic or non-metallic conduit outside of the insulation, except where otherwise protected from the elements, or where out of contact with metal surfaces. Such locations are inside

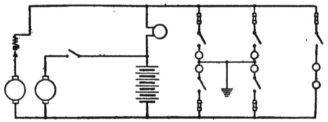

FIG. 304—TWO WIRE SYSTEM WITH SINGLE WIRE FITTINGS

of the body or on wooden dash and instrument boards, or extending in a straight line between chassis and lamps.

The preferred conduit for insulated conductors is a metal tube, either rigid metal or flexible metallic conduit. Where metallic conduit is not used the wires may be run in non-metallic conduit.

Where conductors are run through flexible metallic conduit, the latter must be provided at any open end where wires project, with ferrules having rounded edges.

The fit between the conductor and the ferrule should be sufficiently tight to exclude the free entry of water or oil.

The ends of all conduit where exposed to weather must point downward, so that water will run out of, instead of into, the conduit.

Where metal conduit is led to a connector it must be securely soldered, or permanently clamped to the plug, either inside or outside of a sleeve not less than ¼ inch long.

Where wires are not protected with conduit they are to be cleated at intervals not exceeding 10 inches. Such cleats may be of metal with a layer of insulation interposed between the cleats and wire.

Clearance.—When on wooden dashboard conduit is omitted, no wire shall be nearer to the exhaust pipe than 2 inches.

Where conduit is used the minimum clearance between the exhaust pipe and the conduit shall be 1½ inches.

A minimum clearance of ¼ inch shall be securely maintained between conduit and the nearest carburetor, gasoline pipe, gasoline tank, moving rod and moving lever.

Grounding.—Wherever an insulated conductor is connected to "ground," the connection shall be installed so as to be readily accessible and removable on the completed car.

When the frame of the car is relied on as the ground return, connections shall be made to the chassis frame, or to a substantial part firmly attached to the frame at two or more places, i. e., connections must be to chassis parts in contradistinction to body parts.

Conductors shall not be attached to the frame except by means of terminals soldered to them. The surface on which the terminals are fastened must be clean and free from oxide or paint. The terminals must be fastened to the frame by means of screws or bolts.

Starting circuits should be run independent of "ground" unless a substantial switch is placed in the "ground" connection circuit in an accessible place. The storage battery should be "grounded" on the positive side, and by only one conductor. This "ground" connection should be made in an accessible place by not more than a single screw or nut, or shall be made readily detachable by means of an accessible switch.

Starting Cable.—If starting cable is connected to the frame the connection should be effected in a thorough manner, preferably by riveting a suitable brass or copper plate to the frame, which should be first thoroughly scraped; soldering as well as riveting being recommended. The cable in turn should be sweated to a suitable terminal lug which is then clamped or bolted to said plate.

All "grounds" should be installed so as to be readily accessible and removable on the completed car.

Apparatus intentionally "grounded" in itself shall be so made that for inspection or test the "ground" can be opened without dismantling the apparatus.

Connectors.—All connectors shall be so made that when disconnected the section which is alive will have all live parts recessed 1/16 inch or more below the end of the shell of the connector.

Springs used in connectors shall not be relied on to carry current.

Protection Against Accidental Short Circuits.—All connecting posts on fuse and junction blocks, or on instruments, generators, motors and switches which must necessarily have exposed live parts, shall be so constructed, recessed or installed that an accidental short circuit cannot be effected with screwdrivers, pliers, wrenches or other tools used while making minor repairs or adjustments.

Where live parts might be liable to accidental short circuits to adjacent metal, they must be provided with suitable insulating caps or bushings, or surrounded with a suitable insulating barrier; or the adjacent bare metal liable to be accidentally connected to them through repair tools, should be covered with insulating material.

Battery.—The battery must be installed so that neither acid fumes nor overflow of water or electrolyte will cause serious leakage of current.

Battery hold-down bolts, if connected to metal handles or other metal parts at the level of the top of the jars, must be insulated at at least one of their two ends.

Bulb Sockets.—Bulb sockets must be so made that the continuity of the return path for current is not impaired when the lamp is subjected to vibration.

In lamps, hinges shall not be relied on to carry current between the bulb socket and the battery. The train of connections must either be permanent or so spring-held that contact is maintained tight under vibration.

Protective Devices.—No protective device need be installed in the starting circuit.

The current to all lighting and signal circuits shall be passed through protective devices.

Protective devices shall be on the battery side of switches and junction boxes.

When fuses are used the circuits shall be so arranged that the blowing of a single fuse shall not extinguish all the driving lights, or prevent an inspection lamp being connected to a live circuit.

Size of Wires—The size of wire to use is to a certain extent a matter of judgment. For the starting circuit an exceptionally heavy cable, equivalent to from No. 0 to No. 4 B. & S. gauge, should be used, and the length should be kept

376   FUSES, LIGHTING SWITCHES, ETC.

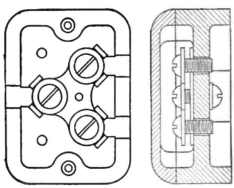

Fig. 307—Junction Box (Westinghouse)

These boxes are mouldings of insulating composition with lids held on by two screws. The connectors are secured to the bases by machine screws screwed in from below and are each provided with two binding screws (three in the case of the junction box) to which the wires leading in and out of the box are fastened.

CHAPTER XXIX

Battery Arrangement—Care and Maintenance of Systems

The capacities of storage batteries employed with starting and lighting systems varies from 360 watt-hours for the smallest to 720 watt-hours for the largest cars. The larger the battery, the less it is affected by the shocks due to the heavy drafts of current in starting, but since the battery is necessarily heavy the capacity is usually chosen as low as practicable. A 6-volt, 60-ampere-hour battery weighs about 45 lbs., and a 6-volt, 120-ampere-hour battery, about 70 lbs. Twelve-volt batteries of the same watt-hour capacity weigh about 20 per cent. more than 6-volt batteries. With a larger battery there is less drop in voltage due to internal resistance and the engine, therefore, will be cranked at a higher speed.

O. W. A. Oetting in an S. A. E. paper on "Effects of Low Temperature on Starting" gave the following battery capacities required with 6-volt systems and a starter gear ratio of about 10 to 1:

Cubic Inches Displacement	Ampere-Hours
Up to 125	70
125 to 200	85
200 to 250	100
250 to 300	120
300 to 350	140
350 to 400	160
400 to 475	180
475 to 600	205

For starting systems of higher voltage or gear ratios greater than 10 to 1, the battery size can be reduced proportionally.

Batteries must be carried in a substantial box at a point not too far from the starter, but particularly where they can be inspected without too much trouble. Formerly the batteries were often carried on the running board in a pressed steel battery box, but fashion now decrees "clean running boards" and the battery is generally hung from the frame, underneath either the front or the rear floorboards. One point that should be kept in view in selecting the location for the battery is that it must be protected against the heat from the engine as much as possible, as otherwise the electrolyte will evaporate very rapidly. Sometimes a shield is provided to deflect the hot air coming from the engine space, and it has been found that this materially reduces the frequency of electrolyte replenishment required.

Mounting of Batteries—Batteries should not be mounted in such a manner that it is necessary to remove them from the car for replenishing the electrolyte. In some instances they have been slung underneath the frame in such a position that they could not be reached from above, and it was neces-

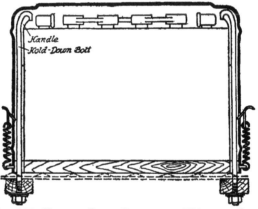

FIG. 307—HOLDING DOWN BATTERY BY MEANS OF BOLTS

sary to loosen studs or bolts holding the battery to the car in order to let it down. This is very poor practice, as if the battery is hard to reach the chances are that it will be neglected and not be refilled until it fails to perform its service, when it has generally been damaged. Accessibility is the chief requirement in locating the battery, as the tendency is to neglect it even if it is accessible.

If the battery is placed in a steel box it is well to provide it with wooden corners, as shown in Fig. 308. These corners serve as guides for the battery and to insure its ventilation. The bottom of the box should be made of wood or the battery should be placed on cleats fastened to the bottom of the metal box. This will serve to keep the bottom

BATTERY ARRANGEMENT

of the battery dry, thus increasing the life of the retainer. Where the box is provided with a wooden bottom, the latter should be grooved and provided with holes, as indicated in the diagram, to allow any overflow to run off. These wooden parts should be boiled in paraffin. In Fig. 308 the box is shown suspended underneath a convenient trap door in the floor of the car.

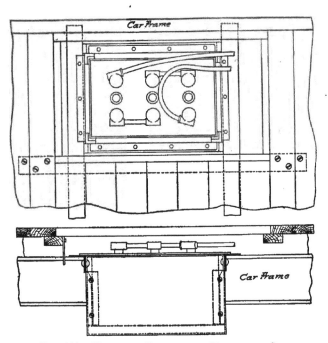

FIG. 308—MOUNTING BATTERY ON PASSENGER CAR

In order that the battery may be held rigidly in position, some means for clamping it down must be provided. Some batteries are now furnished with metal handles bolted to the battery container, the horizontal bar of this handle being provided with a notch adapted to take a hook for holding down the battery, as shown in Fig. 307.

380 BATTERY ARRANGEMENT

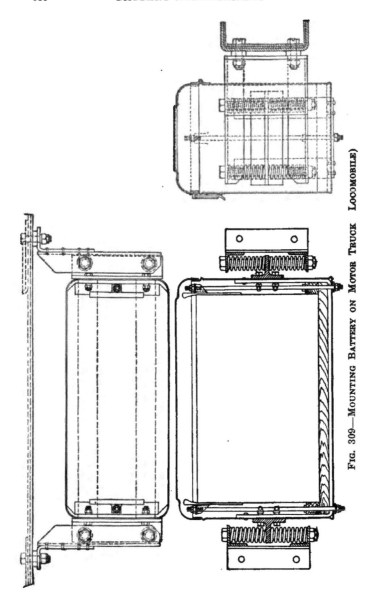

FIG. 309—MOUNTING BATTERY ON MOTOR TRUCK (LOCOMOBILE)

CARE AND MAINTENANCE

Special care must be given to the mounting of the battery on motor trucks, owing to the severe vibration, and some form of auxiliary spring suspension is desirable. This problem has been neatly worked out by the engineering department of the Locomobile Co. of America, whose construction is shown in Fig. 309. Inside a pressed steel box are placed a pair of V-shaped pieces of strip iron, which are bolted to a pair of T section brackets outside the box. The extending arms or webs of these brackets have holes drilled through them and pass over vertical bolts supported from a frame member. There are two coiled springs on each bolt and the bracket rests between these springs.

General Care of Batteries—Owing to the fact that starting and lighting batteries are intended to be automatically kept charged, their care is somewhat different from that which must be bestowed upon batteries which are periodically brought to a charging station. There are chiefly two things to look after: 1, the battery cells must be kept filled with electrolyte and, 2, the battery must be kept charged.

On receiving a battery from the factory or service station, the filling plug should be removed from each cell and a hydrometer test made to see about how much charge the cell holds. If the specific gravity is less than 1.25 the battery should be given a charge from an outside source. If the specific gravity is above 1.25 it is not necessary to charge. If the electrolyte does not cover the plates, either pure water or additional electrolyte should be added. It sometimes happens in shipping the batteries that electrolyte is spilled from the cells and this, of course, must be replaced with additional electrolyte—sulphuric acid solution of about 1.25 specific gravity. On the other hand, if the loss of electrolyte is due to evaporation, it should be made up with rain water or distilled water. If much water or electrolyte has to be added in this manner, the battery, after being refilled, should be given a charge at a low rate, until the specific gravity of the electrolyte of each cell as well as the voltage shows no increase during a period of several hours. After this charge has been completed it is well to take another hydrometer reading of each cell and to correct the specific gravity by adding pure water if it is too high, or electrolyte of 1.30 specific gravity if it is too low, the desired specific gravity in most cases being between 1.27 and 1.285. After replenishing the electrolyte the battery case should be wiped off with cotton waste dipped in ammonia to remove any water or electrolyte that may have been spilled. The battery is then ready for installation in the car.

CARE AND MAINTENANCE

In regular service the height of the electrolyte in the cell should be observed at regular intervals, and the electrolyte replenished if necessary. In regular use it is seldom that anything but distilled or rain water has to be added, as all loss is by evaporation, and only the water evaporates. If one cell regularly shows a greater loss than the rest, it indicates that the cell is leaking and that it must be replaced with a new one. Loss by evaporation, of course, is most rapid in hot weather. After replenishing the electrolyte the filling plugs must be screwed down tightly, for if this is neglected the electrolyte is apt to flow out of the cells, especially during charge.

Making Hydrometer Tests—Generally the brightness of the lamps when the generator is not running gives an indication of the state of charge of the battery, but it is, nevertheless, a good practice to take hydrometer readings frequently so as to ascertain the state of charge of each cell. In making such tests it is well to follow the conventional practice of starting with the cell at the positive end. The tester should always make sure that electrolyte taken from one cell for the purpose of a hydrometer test is returned to that same cell and not to another one. Failure to observe this simple precaution often leads to trouble, as if the electrolyte drawn from one cell is not returned to it, but is later on replaced by distilled water, the specific gravity of the electrolyte in that cell is reduced and the specific gravity of the electrolyte in the other cell is increased, thus disturbing the balance between the different cells. Water can conveniently be added by means of a hydrometer syringe.

Whenever hydrometer readings show the battery to be in a low state of charge, it should be charged either by running the engine while the car is standing still or by taking the battery to a charging station. Hydrometer readings should not be taken immediately after water has been added to a cell, as the solution is then not of uniform density. Ordinarily hydrometer readings are taken before water is added and after the battery has been charged.

Batteries Not Holding Charge—One very frequent trouble with the storage batteries of lighting and starting systems is that they do not remain in a state of charge or will not "hold the charge." This is due to a variety of causes. In the first place it is possible that the car is being driven under such conditions that the generator very seldom runs at charging speeds, whereas a great deal of current may be taken from the battery unless the charging rate of the system is adjustable. There is no other remedy for this trouble than to change the conditions of operation, that is, drive the car more at high speed, make fewer starts and use the lights more sparingly. One common cause of batteries refusing to hold the charge is that they have been neglected and allowed to sulphate. This often occurs when the batteries are first put into service. If a considerable interval of time elapses after they are shipped from the factory until they go into service on the car, they are in a low state of charge, and unless they are brought up to full charge before being placed in regular service on a car they are likely to give trouble at an early date. Sometimes cars with starting and lighting outfits are kept in a dealer's show room for several weeks and even months, the battery being allowed to run down, and when a car is delivered to a customer the battery is in bad condition. When a battery has become incapacitated by having been allowed to remain in a discharged condition for a considerable period of time, the only way to restore it to a condition of serviceability, is to give it a long continued charge at a very low rate. During this charge frequent hydrometer readings are taken, and the specific gravity of the electrolyte will sometimes continue to increase for several days. As long as the gravity increases the charge is continued. In most ordinary cases a battery can be rejuvenated in two or three days, but in aggravated cases it will take as long as fourteen days to get a battery back to its normal capacity.

Care of Storage Battery in Winter—Many cars equipped with starting and lighting systems are placed out of commission during the winter season and care must then be taken not to let the storage battery deteriorate during this period. The battery gradually loses its charge by leakage, and as the sulphate formed on the battery plates during discharge becomes hardened if the battery is allowed to stand in the discharged condition, it must be given a charge at regular intervals. If the car is stored in a place where it is readily accessible, perhaps the best plan is to run the engine once every two weeks for a sufficient length of time to bring the battery up to full charge as indicated by hydrometer test. In most cases one hour's running every two weeks will be sufficient to keep the battery in good condition.

If the car is stored so it cannot readily be gotten at, it is well to remove the battery from the car, and to either charge it from house mains by means of an installation as described in connection with ignition batteries, or else bring it to a charging station. If the battery is charged through a lamp resistance from direct current mains or through a rectifier from alternating current mains, a charge over night or for about 12 hours every two weeks will be sufficient to keep it in good condition. Before every such charge the height of the electrolyte should be observed, and be corrected if necessary.

One reason for keeping storage batteries in a fair state of charge during periods of idleness in winter time is that a dilute solution of sulphuric acid as found in a discharged battery freezes at about zero Fahr., whereas a solution of the density found in a fully charged battery never freezes at any temperature encountered in this latitude. A curve of freezing temperatures of sulphuric acid solutions of different densities is shown in Fig. 310. To prevent injury from freezing, it is advisable after water is added to a battery in cold weather, to immediately start the engine and charge the battery for a while, for, since the water is of lower specific gravity than the solution, any water added will remain at the top of the cell and may freeze if the temperature is sufficiently low. To obviate any chance of trouble from freezing it is best to store the batteries in a room in which the temperature does not fall below 40 degrees Fahrenheit. The room should also be dry so as to minimize leakage of current.

In describing the different elements of lighting and starting systems most of the ailments which these elements are heir to have been enumerated and their symptoms and remedies pointed out. It will be well, however, now to take the electric system as a whole and consider various symptoms of trouble and the location and elimination of these troubles.

Lamps Burn Dimly—The most frequent trouble in operating an electric lighting system is probably that the lamps, upon being switched on, burn dimly. If this occurs while the engine is at rest so that the lamps can draw current only from the battery, it shows that the battery is in a low state of charge. This may be due either to natural causes, such as heavy consumption of current in consequence of extensive night driving and little opportunity for charging, or it may be due to defects in the battery, the generator, the regulating or wiring system. If there is nothing else wrong than that the battery has been carelessly allowed to become discharged, then as soon as the engine is started and run up to a speed where the automatic battery switch closes the circuit, the lights will flash up brightly.

The symptoms would be the same, however, if the battery, instead of being discharged due to unfavorable conditions of car operation, had become defective in some way. For instance, if the electrolyte had been allowed to evaporate, so that, say, only one-half of the plates was covered, or if the battery had been allowed to stand in the discharged condition so that its plates had become covered with sulphate, then the capacity would be greatly reduced, as would be the charging rate with some systems of control. However, if the lights burned dimly for the first time it would be safe to assume that nothing was wrong with the battery other than that it needed a charge, which should be given it as soon as conditions permit.

No Current from Generator—Now assume that when the

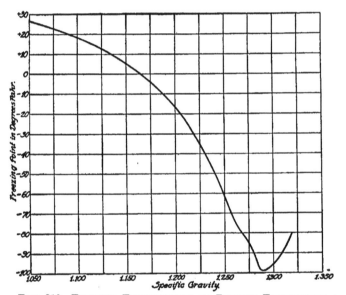

FIGS 310—FREEZING TEMPERATURES OF BATTERY ELECTROLYTE

engine is started and run up to speed there is no change in the brightness of the lamps. Then, evidently, no current from the generator reaches the lamps. This may be due either to failure of the generator to generate, to a break in the circuit somewhere or to a short circuit.

In such a case the best plan is to first investigate whether or not the generator is generating, and for this purpose a low scale voltmeter is very handy. Applied to the terminals or the brushholders of the generator it will indicate not only whether the generator is generating, but the exact voltage it is producing. If no voltmeter is available the terminals of an inspection lamp may be applied to the brushes and the lamp will light up if the generator generates.

If it is found that the generator works properly, and yet no current reaches the lamps, it can only be due to failure of the automatic switch to close the circuit or to some defect in the wiring. The automatic switch should first be examined. This, if of the electromagnetic type, is operated by a circuit connected in shunt to the generator. If this circuit should be broken at any point no current could flow, and the switch could not act. The circuit would most likely be interrupted at one of the connections, which may have shaken loose, and it would be a good plan to examine all binding posts and connectors for looseness. If this does not disclose the defect, and if it has been shown that the generator is working properly, see whether the core of the automatic switch is energized, which can be done by bringing a piece of soft iron or steel, such as a knife blade, near it. If there is no magnetic attraction on the knife blade, etc., then no current flows through the switch coil and a further search for a break in the circuit (or possibly a short circuit) should be made. On the other hand, if the test shows the core of the switch to be magnetic and yet the generator main circuit is not closed, then the defect must be mechanical; that is, the switch arm must not be free to move. It should be examined and any fault corrected.

Generator not Generating—Now suppose that the test of the generator showed that it was not producing current, then the next thing to do would be to locate the cause of this trouble. That, of course, must be inside the generator itself. The natural thing to do would be to first suspect the current carrying wearing parts, that is, the commutator and brushes. There may be poor contact between the brushes and the commutator which may be due to (1) dirt or oil on the commutator, (2) brushes stuck in brushholder, (3) brushes worn too short, (4) wearing surface of commutator rough on account of loose segments or projecting mica plates, (5) brush spring broken or displaced.

First examine the brushes that they are free in the brushholders and are pressed against the commutator with adequate force. It should be borne in mind that the brushes perform a very important part in the operation of the generator and that not any kind of brush will do. The material must have low electrical resistance, a low coefficient of friction, must not be so hard as to wear the commutator out too quickly nor so soft as to wear itself excessively. In purchasing new brushes care should therefore be exercised to see that they are of good quality.

It has already been pointed out that when the commutator shows signs of roughness it should be smoothened with fine sandpaper (not emery cloth) by a method described in a previous chapter. Lubricant should be applied to the commutator sparingly, if at all. A little lubricant is of benefit, as it prevents squeaking and cutting, but inasmuch as "green" operators are very apt to use lubricant to excess, and as this causes no end of trouble, some makers advise against the use of any lubricant whatever. A little machine oil may be put on with the tip of the finger, or a stick of commutator compound may be lightly applied to the surface of the revolving commutator. If deep grooves have worn into the commutator, or if it is otherwise very rough, it should be turned down in a lathe.

Electrical Defects in Generator—If the commutator on inspection shows no mechanical defects, or if after the commutator has been smoothened and cleaned the generator still fails to generate, the chances are that the trouble is due to some electrical defect either in the commutator or armature. One or more of the commutator segments may be grounded or short circuited, armature leads may have become disconnected from the commutator lugs, or the armature coils may be grounded or short circuited. A ground can easily be detected by applying test points as shown in Fig. 312 to a portion of the armature winding or a commutator segment and to the armature shaft respectively. If the lamp lights up there is a ground somewhere on the armature or commutator. Short circuits in the commutator can be detected by means of test points only after all the armature leads have been disconnected, the same method being used as for grounds, except that the ends of the wire are always applied to adjacent commutator bars.

Tests Points—The test points illustrated in Fig. 312 are readily made by any mechanic. The points proper consist of ¼-inch or 5/16-inch steel rods with one end sharpened to a point. A small hole is drilled centrally into the other end and into this the end of the wire is soldered. This wire is preferably lamp cord, and the wires from the test points are connected into an incandescent lamp circuit, as shown. Short lengths of fiber tube are forced over the test points. They serve as convenient insulating grips and also prevent the wires from breaking where they join to the steel rods.

Such test points are of great service in testing out electric circuits and apparatus. If the lamp lights up when the points are touched to any two metallic parts, these parts are in metallic connection, or, in other words, the circuit between them is complete. If the lamp does not light up, then there is no circuit between the parts touched. In this way tests can be made for open circuits, grounds and short circuits.

Some electricans prefer an audible to a visible signal and make use of a testing buzzer. The outfit consists of a wooden box containing two dry cells connected in series, with a buzzer mounted on top of the box. For quickly testing a large number of parts in bright daylight the buzzer is probably better than the lamp, as the lighting up of the lamp then is not so noticeable.

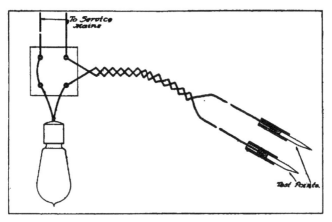

Fig. 312—Testing Points and Their Method of Connection

However, if a milli-voltmeter is available "shorts" in the commutator or armature coils can be located without disconnecting the armature leads from the commutator. The method is as follows (Fig. 313): Apply the wires from a dry cell to diametrically opposite commutator bars. Then apply the milli-voltmeter points to adjacent commutator bars successively all around the commutator. The voltmeter readings between all pairs of adjacent commutator bars should be the same. If the voltmeter shows no voltage drop between any pair of bars, then there is a short circuit between them; if the reading is less than that between the other pairs of bars, then part of the armature coil between these bars is short circuited. If the milli-voltmeter shows no reading between a number of pairs of adjacent bars and the full voltage of the cell between another pair of bars, then the coil between these latter bars is open.

If the armature has become grounded or burned out (which latter defect can easily be detected by a careful inspection of the windings) it must be sent to an electrical repair shop for rewinding.

Open Field Circuit—Failure to generate may also be due to a break in the main field circuit. To determine whether the field circuit is complete, open the armature circuit by removing one of the brushes and run wires from the battery or from the battery connections on the junction box to the ter-

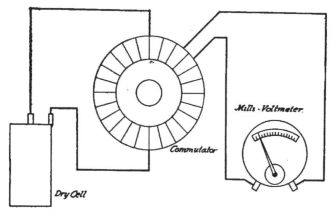

FIG. 313—TESTING ARMATURE COILS FOR "SHORTS"

minals of the shunt winding. If the shunt circuit is complete a bright spark is produced when the circuit is broken, owing to the self-induction of the field circuit.

Some users of systems comprising no ammeter or charging indicator, in order to determine whether the generator is generating, disconnect one of the battery leads and see whether the lights keep on burning. While this practice is permissible with some systems, it is inadvisable with others in which the storage battery constitutes the voltage control element, as with such systems the voltage is likely to rise beyond the normal voltage of the bulbs and burn them out. If the test is made at all it should be made with the engine running at a speed only slightly above that at which charging begins.

It is obvious that when all lamps burn dimly or not at all the trouble must be with the source of current or somewhere in the main line, whereas if defective operation is limited to a single consuming device or set of consuming devices the cause of trouble must be either in them or in their individual circuits. This consideration sometimes helps in locating troubles. If a single lamp fails to burn while all the others light up properly the filament of the bulb is probably broken or there may not be good electric contact in the socket. Short circuits and grounds in the wiring, if the installation is made in accordance with the S. A. E. recommendations, printed elsewhere, are extremely unlikely, as the insulation of the cables is fully protected both against mechanical injury and against oil and other solvents of rubber. However, there are some cars running in which the wiring is not carried in protecting tubing as required by the S. A. E. specifications, but is subject to the destructive action of oil accumulations, to abrasion due to insecure support and to accidental injury by tools, etc. In such cases short circuits and grounds are very apt to occur, especially after the car gets old, and are a prolific cause of trouble. If any of the wiring is in a part of the car where oil is constantly leaking, the insulation will rot off, and if it is replaced it is best to use wire having a protective braiding or coat of enamel, or else the wire should be enclosed in circular loom at this point.

It is never a good plan to waste current unnecessarily, as by burning the headlights while the car is at a standstill, because it is not as bad to overcharge the battery as to undercharge it, and in most systems and under most conditions of operation there is a tendency to undercharge. One practice that is to be strongly deprecated is to start the car or run the car on the electric starter. It is obvious that more current is drawn from the battery by the starter if the latter, instead of merely starting the engine, must start the car besides. The service of a starting battery is quite severe enough under the most favorable conditions, and there is no sense in aggravating it in any way. In starting, if the engine fails to pick up its cycle after it has been cranked a reasonable time, it is foolish to continue cranking, as this will result only in the exhaustion of the battery. In such a case the starter should be stopped and the fuel and ignition systems be examined with a view to locating the cause of the trouble.

APPENDIX

SPECIFICATIONS FOR HEADLIGHT TESTS*

For the purpose of test the intent of the New York State law dealing with automobile headlights and providing that front lights shall be so arranged, adjusted and operated as to avoid dangerous glare or dazzle, and so that no dangerous or dazzling light, projected to the left of the axis of the vehicle when measured 75 feet or more ahead of the lamps, shall rise above 42 inches on the level surface on which the vehicle stands, such front lights shall be sufficient to reveal any person, vehicle or substantial object on the road straight ahead of such motor vehicle for a distance of at least 200 feet, is deemed to be complied with if the following conditions are fulfilled:

1. Any pair of head lamps under the conditions of use must produce light which, when measured on a level surface on which the vehicle stands at a distance of 200 feet directly in front of the car and at some point between the said level surface and a point 42 inches above this surface, is not less than 1200 apparent candlepower.

2. Any pair of head lamps under the conditions of use shall produce light which, when measured at a distance of 100 feet directly in front of the car, and at a height of 60 inches above the level surface, on which the vehicle stands, does not exceed 2400 apparent candlepower, nor shall this value be exceeded at a greater height than 60 inches.

3. Any pair of head lamps under the conditions of use shall produce light which, when measured at a distance of 100 feet ahead of the car, and 7 feet or more to the left of the axis of the same, and at a height of 60 inches or more above the level surface, on which the vehicle stands, does not exceed 800 apparent candlepower.

Conditions of Laboratory Test—In order to determine whether any particular device conforms to these requirements, it shall be subjected to laboratory tests according to the following specifications:

Number of Samples.—Two pairs of samples of the device submitted shall be subjected to test. In the case of front glasses, the sample shall be of 9¼ inch diameter, when practicable.

---

*As required under the New York State vehicle lighting law.

APPENDIX

Reflectors and Incandescent Lamps—The reflectors used in connection with the laboratory tests shall be of standard high-grade manufacture of 1.25 inch focal length, with clean and highly polished surfaces and as nearly truly paraboloidal in form as practicable, and as approved for this purpose by the National Bureau of Standards.

The incandescent lamps used in connection with the laboratory test shall be of standard high-grade manufacture and as approved for this purpose by the National Bureau of Standards.

Adjustments by Manufacturers' Representative—The manufacturer of the device shall be given due notice of the date and place of test. Manufacturers' representatives present at the test shall be privileged to adjust their devices in any way which represents an ordinary and legitimate adjustment, including tilting the lamps or reflectors, which can be carried out by purchasers of the device, or such adjustment may be made by the laboratory expert acting on the instructions of the manufacturer. The character of the adjustments so made shall be carefully noted and stated in the report as manufacturer's adjustment.

Tests—The tests shall be as follows:

Test 1. Four-point test of pairs of samples.

A pair of testing reflectors, mounted similarly to the head lamps on a car, shall be set up in a dark room at a distance of not less than 60 feet nor more than 100 feet from a vertical white screen. If a testing distance of 100 feet is taken, the reflectors shall be set 28 inches apart from center to center, and if a shorter testing distance is taken, the distance between reflectors shall be proportionately reduced. The axes of the lamps shall be parallel and horizontal, or tilted in accordance with the manufacturers' adjustment. The intensity of the combined light shall then be measured with each pair of samples in turn, with the reflectors fitted with a pair of each of the following types of incandescent lamps, in turn.

(1) Vacuum type, 6-8 volts, 17 mscp., G-12 bulb.

(2) Gas-filled type, 6-8 volts, 20 mscp., G-12 bulb.

The lamps shall be adjusted to give their rated candlepower. Measurements shall be made at the following points at the surface of the screen:

A. In the median vertical plane parallel to the lamp axes, on a level with the lamps.

B. In the same plane one degree of arc below the level of the lamps.

C. In the same plane one degree of arc above the level of the lamps.

D. Four degrees of arc to the left of this plane and one degree of arc above the level.

In an acceptable device both pairs of samples shall conform to the following specifications for observed apparent candlepower.

Points A and B. At at least one of these points the apparent candlepower shall not be less than 1200.

Point C. The apparent candlepower shall not exceed 2400.

Point D. The apparent candlepower shall not exceed 800.

Provided, however, that if the test indicates that a device which is unacceptable with either of the test lamps will come within the specifications with lamps of another candlepower or of the other type, the device may be passed with corresponding limitations as to the incandescent lamps to be used in connection with it. Test 2. Complete test of single sample.

A single sample taken as an average representative of the device as manufactured, shall be submitted to a complete test with a vacuum incandescent lamp of 17 candlepower, 6-8 volt rating in a G-12 bulb. This test shall show its light distribution characteristic by actual measurements made according to recognized and exact methods.

Distribution of Samples—One pair of the samples submitted shall be retained by the testing laboratory for purpose of future reference and as sample of construction, and the other pair shall be returned to the office of the Secretary of State.

Report—The report of the tests shall be rendered in duplicate to the Secretary of State, and shall be signed or initialed not only by the expert making the test, but also by an executive officer of the institution making the test.

It shall include a statement by the testing laboratory as to whether the device when properly applied substantially complied with Section 286 of the Highway Law and shall suggest the maximum candlepower to be used with the same, and as to the other conditions necessary in the operation of the device, in such a way that it will comply with the requirements of this specification.

Head Lamp Lighting Nomenclature (S. A. E.)

Light refers to the luminous radiation emitted from the head lamp, and may be divided as follows:

Direct Light is the light emitted directly from the filament and emerging from the head lamp.

Stray Direct Light is the light unintentionally changed in direction by the reflection from bulb parts.

Direct Reflected Light is that part of the light emerging from the head lamp which has been specularly reflected by the reflector in the form of a defined beam of light.

Stray Reflected Light is Stray Direct Light which has been specularly reflected by the reflector.

Diffused Reflected Light is that part of the emitted light which has been diffusively scattered by the reflector owing to imperfections of the reflecting surface.

394 APPENDIX

Light Cone is the solid angle of light emitted in the form of a cone by the head lamp, and is subdivided for reference as follows:

Direct Cone is the direct and stray direct light from the incandescent lamp passing through the front glass.

Direct Reflected Beam Cone is that cone of light formed by the blending of all cones which are produced by the reflected images of the filament and which cones emanate from all points of the reflector upon which light falls. The cone produced in this manner includes only the direct reflected light.

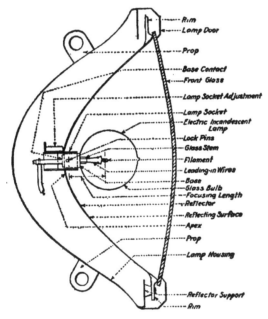

FIG. 314—HEADLAMP NOMENCLATURE

Scattered Light Cone is that cone of light formed by the blending of all direct, stray direct, stray reflected and diffused reflected light passing through the front glass.

The **Lamp Door** is the front cover of the Lamp Housing, regardless of the method of attachment.

The **Lamp Housing** is that portion of the lighting units which forms the exterior shell or case.

The **Lamp Socket** is the receptacle which receives the electric incandescent lamp base.

APPENDIX

The **Lamp Socket Adjustment** is the device for regulating the position of the Lamp Socket with reference to the reflector.

The **Prop or Props** are that portion of the Lamp Housings by means of which a lighting unit is fastened to its supporting bracket.

(a) The **Reflector Apex** is that portion of the reflector to the extreme rear.

(b) The **Reflector Rim** is that portion which retains the reflector in place and which forms no part of the reflecting surface.

(c) The **Reflecting Surface** is that portion of the reflector between the apex and the rim which receives and redirects rays of light.

The **Reflector Support** is the means by which the reflector is retained in a permanent position.

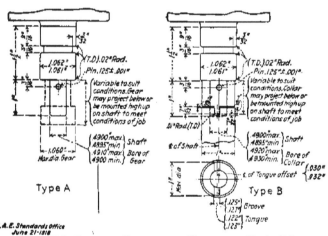

Fig. 315—Ignition Distributor Mounting (S. A. E.)

Electric Incandescent Lamp Voltage (S. A. E.)

Number of storage battery cells in system..	3	4	6	9
Nominal voltage of electric incandescent lamps	6–8	8–10	12–16	18–24

Bulb Sizes (S. A. E.)

Spherical bulbs used in head lamps shall be no smaller than 1½ in. diameter nor larger than 2 1/16 in. diameter.

Spherical bulbs used in side lamps shall be no larger than 1 in. diameter.

Spherical bulbs used in rear and instrument lamps shall be **no** larger than ¾ in. diameter.

APPENDIX

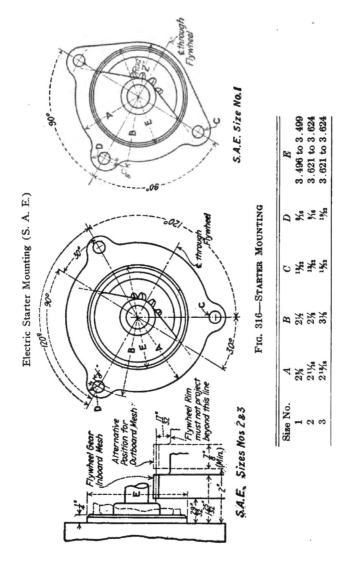

Fig. 316—Starter Mounting

Size No.	A	B	C	D	E
1	2⅞	2⅞	13/32	9/16	3.496 to 3.499
2	2 11/16	2⅞	13/32	9/16	3.621 to 3.624
3	2 11/16	3⅛	13/32	19/32	3.621 to 3.624

Insulated Cable for Gasoline Cars (S. A. E.)

Class A—Rubber-Compound Insulation (Secondary Cable)—This insulation shall consist of a vulcanized rubber compound, not previously used, containing not less than 30 per cent by weight of good Hevea rubber that will withstand the tests outlined in the following paragraphs.

Physical Tests—A test-specimen of rubber compound having a length of not less than 6 in., taken from a cable, shall have marks placed upon it 2 in. apart. The sample shall then be stretched at the rate of 12 in. per minute until these marks are 6 in. apart; the test-specimen shall then be released within 5 sec. and the measurement taken 1 min. thereafter, when the distance between these marks shall not exceed 2⅜ in. The test-specimen shall then be stretched until the marks are 9 in. apart before rupture. This test shall be made at a temperature of not less than 10 deg. C. (50 deg. Fahr.).

The ultimate tensile strength of the rubber compound shall not be less than 1000 lb. per square inch, calculated upon the original cross-section of test-specimens before stretching. This test shall be made at a temperature of not less than 10 deg. C. (50 deg. Fahr.).

Electrical Tests—Each specimen of rubber insulated cable shall successfully withstand a voltage test of 12,000 A. C. for 5 min. after 12 hr. submersion in water and while still immersed. After the voltage test the cable, while still immersed, shall have an insulation resistance of 2500 megohms per mile at 15.5 deg. C. (60 deg. Fahr.) after 1 min. electrification. These two tests shall be made in accordance with the Standardization Rules of the American Institute of Electrical Engineers.

Any 1-ft. sample of this rubber insulated cable must show a dielectric strength sufficient to resist for 5 min. the application of 20,000 volts A. C. This test shall be made as specified in the National Electric Code.

Life Test of Cable—The following aging test should be made to determine the life of the cable. Take a length of cable, preferably 5 ft., remove insulation from both ends, solder one end of the cable to the end of a steel rod of same diameter as outside diameter of cable, wrap the cable tightly around this rod and solder the other end of cable to the rod so that it will retain its tightly wrapped condition. Leave exposed in this condition to the elements. If cable is not properly compounded and properly vulcanized, deteriorating effects, such as hardening, becoming brittle and cracking, will be apparent in from one to three months.

Sizes with Cotton-Braid Covering—Secondary cable is to be made in two sizes with overall diameters of 7 and 9 mm. for plain rubber insulation. If a varnished cotton-braid is desired the same dimensions are to be used with the addition of glazed cotton braid approximately 1/64 in. thick making the overall diameters approximately 5/16 and 25/64 in. The braid is to be treated with at least two coats of insulating varnish dried separately.

Rubber-Compound Insulated Secondary Cable

Nominal Size	Number of Wires	Diameter of Wires	Circular Mils in Cable
7 mm.	26	.010 in.	2600
9 mm.	37	.010 in.	3700

Nominal Size	Diameter Over Wire	Diameter Over Rubber (Max.)	Thickness of Rubber (Min.)
7 mm.	.065 in.	.280 in.	.100 in.
9 mm.	.072 in.	.360 in.	.125 in.

Class B—Rubber-Compound and Fabric Insulation (Primary Cable)—This insulation shall consist of a rubber compound and two thicknesses of fabric. The compound is to be evenly applied in thickness and must conform to the physical and electrical requirements of the National Electric Code (1915 edition).

The rubber compound is to be covered first either with an overlapping strip of varnished cambric .008 to .010 in. thick, or a cotton braid at least 1/64 in. thick, so saturated as to make it oil- and moisture-proof.

Cable No.	No. of Wires	Dia. of Wires, In.	Nominal Gage of Wires	Circular Mils in Cable
14	41	.0100	30	4,100
12	49	.0113	29	6,208
10	49	.0142	27	9,854
2	133	.0226	23	67,764
1	133	.0254	22	85,466
0	133	.0285	21	107,743
00	133	.0320	20	135,926

Cable No.	Carrying Capacity, Amp.	Diameter Over Wire, In.	Diameter Over Braid (Max.), In.	Thickness Rubber (Min.), In.
14	....	.080	.22	1/32
12	....	.102	.25	1/32
10	....	.128	.27	1/32
2	92	.340	.53	3/64
1	102	.380	.60	1/16
0	127	.427	.65	1/16
00	150	.480	.71	1/16

Stranding: Number 14 to be bunched; numbers 12 and 10 to be bunched or rope lay; numbers 2, 1, 0, 00 to be rope lay.

APPENDIX

The first fabric is to be covered with an outer braid of strong protective character and at least 1/64 in. thick, preferably glazed and treated with at least two coats of insulating varnish dried separately.

Test Voltage—The completed cable shall be capable of withstanding for 1 min. 1000 volts A. C. applied between the copper conductor and a metal foil wrapped around the outside of the insulation. The frequency of the test circuit shall be in accordance with the Standardization Rules of the American Institute of Electrical Engineers.

Class C—Varnished Fabric Insulation (Primary Cable)—This insulation shall consist of either two or three layers of varnished cambric tape as specified in the table below, each layer from .008 to .010 in. thick.

The tape shall be covered with one closely woven cotton braid at least 1/64 in. thick, so treated as to be oil- and moisture-proof and with one closely woven outer braid at least 1/64 in. thick, of good strong protective character, preferably glazed, and treated with at least two coats of insulating varnish dried separately.

Fabric Insulated Primary Cable

Cable No.	Number of Wires	Diameter of Wires, Inches	Nominal Gage of Wires	Circular Mils in Cable
14	41	.0100	30	4,100
12	49	.0113	29	6,208
10	49	.0142	27	9,854
2	133	.0226	23	67,764
1	133	.0254	22	85,466
0	133	.0285	21	107,743
00	133	.0320	20	135,926

Cable No.	Carrying Cap. Amp.	Dia. Over Wire, In.	Dia. Over Braid (Max.) In.	Layers of Cambric
14	....	.080	.200	2
12	....	.102	.230	2
10	....	.128	.250	2
2	127	.340	.478	3
1	152	.380	.518	3
0	202	.427	.568	3
00	227	.480	.628	3

Stranding: Number 14 to be bunched; numbers 12 and 10 to be bunched or rope lay; numbers 2, 1, 0, 00 to be rope lay.

Test Voltage—The completed cable shall be capable of withstanding for 1 min. a test voltage of 1500 A. C. applied between the copper conductor and a metal foil wrapped around the outside cable. The frequency of the test circuit shall be in accordance with the Standardization Rules of the American Institute of Electrical Engineers.

Conductors—The conductivity of the individual wires in a strand shall conform to the table as given in the National Electric Code (1915 edition). All wires are to be thoroughly annealed and tinned.

Test for Tinning—The following test on tinning as adopted by the Railway Signal Association is recommended:

"Samples of wire shall be thoroughly cleaned with alcohol and immersed in hydrochloric acid of specific gravity 1.088 for 1 min. They shall then be rinsed in clear water and immersed in a solution of sodium sulphide of specific gravity 1.142 for 32 sec. and again washed. This operation* should be repeated three times and if the samples do not become clearly blackened after the fourth immersion, the tinning shall be regarded as satisfactory. The sodium sulphide solution must contain an excess of sulphur and should have sufficient strength to thoroughly blacken a piece of clean untinned copper wire in 5 seconds."

Flexible Steel Tubing for Automobile Electric Wiring (S. A. E.)

The accompanying table relates to packed tubing, but the same outside diameters should exist in unpacked tubing. It is considered desirable that packing should be used for the sake of tightness. Cotton packing can be used where a fireproof material is not required.

Sizes of Flexible Steel Tubing

Inside Diameter	Max. Outside Diameter	Min. Thickness of Strip	Inside Diameter	Max. Outside Diameter	Min. Thickness of Strip
3/16	17/64	0.010	9/16	11/16	0.014
7/32	19/64	.010	5/8	3/4	.014
1/4	11/32	.010	11/16	27/32	.014
9/32	3/8	.012	3/4	29/32	.016
5/16	13/32	.012	13/16	31/32	.016
3/8	15/32	.012	7/8	1 1/32	.016
7/16	35/64	.012	1	1 5/32	.016
1/2	39/64	.014			

All dimensions in inches.

Width of strip and depth of flange are left optional with the manufacturer, so long as the required degree of flexibility and the specified inside and outside diameters are obtained.

*"This operation" is interpreted as meaning the immersion in both hydrochloric acid and the sodium sulphide solution.

Where galvanized strip is used, it should be electrically galvanized and the amount of zinc should not be less than 0.06 oz. per sq. ft. of surface.

Test for Galvanizing of Tubing—The copper-sulphate method is recommended for testing the galvanizing of flexible tubing. This cannot be considered as an exact test, but it does not require a chemist. With experience, anyone can determine quickly by comparison the amount of zinc coating.

The test solution is to be prepared by boiling pure water with sufficient crystals of chemically-pure copper sulphate to give a saturated solution with an excess of copper sulphate when cool. The solution shall be neutralized by the addition of oxide of copper, then filtered and diluted to a specific gravity of 1.186 at 65 deg. F. (A saturated solution requires about 27 parts of copper sulphate to 73 parts of water, by weight.)

In order to guard against changes in specific gravity or saturation, the solution should not be kept in a room whose temperature falls much below 65 deg.

Method of Making Galvanizing Test—The solution shall be poured into a No. 1 tall glass beaker about 2 in. diameter; about 100 cu. cm. shall be used for test of ¾-in. and smaller conduits, and about 125 cu. cm. for larger sizes. The same supply of solution shall be used for successive dips of any one sample, but a fresh supply shall be used for each new sample, and the used solution shall not be returned to the supply bottle.

The solution in the beaker shall be brought to a temperature of approximately 65 deg. F. before the test is made, by placing the beaker in a larger vessel containing warmer or colder water and stirring with a glass thermometer.

Pieces of conduit about 6 in. long shall be used for the test, with the ends cut off approximately square, cooled before the test if heated by sawing. Each sample shall be washed in running water, dipped up and down in a vessel containing either carbon tetrachloride or ether, and allowed to dry before being placed in the test beaker.

The sample as thus prepared shall be stood on end in the copper-sulphate solution for one minute, without moving or stirring the solution. At the end of one minute the sample is to be removed, rinsed in running water, and wiped lightly, inside and outside, with white cheese-cloth until dry. Violent rubbing and contact with the hand or any other object are to be avoided.

Interpretation of Test—The zinc coating of the sample is dissolved by the solution and a coating of copper deposited in its place. As long as zinc remains the copper can be easily removed by washing and wiping, but when the zinc has been entirely dissolved at any point the copper deposited at that point will not wash or wipe off. The number of dips required to cause a fixed copper deposit is, therefore, in some degree, a measure of the thickness of the zinc coating.

APPENDIX

A sample of flexible tubing shall pass this test if it does not have more than 25 per cent of its immersed surface covered with a fixed copper deposit after one immersion for one minute. Several samples should be tested and a comparison noted.

Each sample shall be subjected to at least one additional dip, and the dips are generally continued until the entire immersed surface is coated with a fixed copper deposit.

Dimensions of Lead Batteries for Lighting and for Combined Lighting and Starting Service (S. A. E.)

The overall width of the battery, measured from side to side of case, shall not exceed 7½ in.

The overall height of the battery measured from bottom of case to top of handles shall not exceed 9½ in.

The overall length of the battery, measured from end to end of case, including handles, shall vary according to the capacity of the battery and its details of design. Handles shall, as standard, be placed at the ends of the battery, and provision for hold-down devices shall, as standard, be made at the ends of the battery. The space occupied by such handles and hold-down devices shall be in the direction of the length of the battery only, and not in the direction of its width. Terminals and connections shall not extend above the handles, the latter to be the higher point.

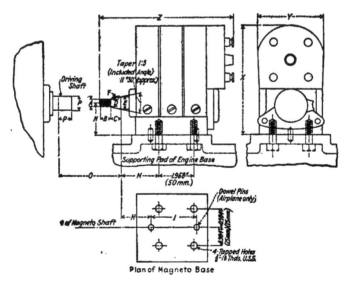

FIG. 317—STANDARD MAGNETO DIMENSIONS (S. A. E.)

	*4 and 6 Cyl. †Airplane, Auto, Marine Tractor		8 and 12 Cyl. †Airplane Automobile		Motorcycle	
	Inch	Mm.	Inch	Mm.	Inch	Mm.
A	3/8"–16 thds. U.S.S.	9.53–1.58 Mm. pitch U.S.S.	3/8"–16 thds. U.S.S.	9.53–1.58 Mm. pitch U.S.S.	5/16"–18 thds. U.S.S.	7.94–1.41 Mm. pitch U.S.S.
B	.5905	15	.5905	15	.4170	10.59
C	.590	14.98	.590	14.98	.3543	9
D	.472	12	.472	12	.4409	11.20
E	.590	14.98	.590	14.98	.5118	13
Woodruff Key F	No. 3 1/2x1/4x1/8	No. 3 12.7x 6.35x 3.18	No. 3 1/2x1/4x1/8	No. 3 12.7x 6.35x 3.18	Special 5/16x5/32x 3/32 thick	Special 7.94x 3.97x 2.38
	Proj. 1/16	Proj. 1.59	Proj. 1/16	Proj. 1.59	Proj. 1/16	Proj. 1.59
H	No std.	.....	1.693	43	No std.	.....
I	No std.	.....	2.519	63.98	No std.	.....
M	2.086	52.98	2.086	52.98	1.553	39.45
N	1.771	44.98	1.968	50	1.771	44.98
O	2.375	60.32	No std.	.....	No std.	.....
P	0.75	19.05	No std.	.....	No std.	.....
P (dia.)	0.749 0.750	.....	.....	.....	.....	.....
Magneto Space (Max.) X	8.000	203.20	9.000	228.60	6.000	152.40
Magneto Space (Max.) Y	5.000	127.00	5.000	127.00	3.750	95.25
Magneto Space (Max.) Z	10.000	254.00	10.000	254.00	6.375	161.93
Width at Brushes—W	No std.	.....	No std.	.....	4.000	101.60
Advance Lever Radius	2.125	53.97	No std.	.....	1.968	50
Timing Lever: Tapped Hole	1/4"–28 thds. S.A.E.	6.35–.9 Mm. pitch S.A.E.	No std.	.....	No std.	.....
Timing Lever: Plain Hole	.25	6.35	No std.	.....	.2185	5.55
Dowels (Airplane Only): Diam.	No std.	.....	5/16 +.002 −.000	7.94 +.05 −.00	No std.	.....
Dowels (Airplane Only): Depth	No std.	.....	5/16	7.94	No std.	.....

Insulation Requirements of Electrical Apparatus After Installation on Gasoline Automobiles (S. A. E.)

Electrical apparatus for use on gasoline automobiles, when operated on circuits of from 6 to 25 volts, shall be capable after installation of withstanding for one minute an alternating potential of 500 volts, the test being applied between the conducting circuit and frame or ground. In the case of apparatus with one terminal grounded, the ground connection shall be removed at such a point as will permit the test being applied to all parts of the circuit which, in actual use, will be subjected to working potential.

Exception.—Batteries will not be subjected to any insulation test above their working potential.

Tail-Lamp Glasses (S. A. E.)

The standard tail-lamp glass is 3 in. diameter with tolerances of plus 1/32 in. and minus 1/64 in.

Head-Lamp Socket—Focusing (S. A. E.)

The position of the slots in the head-lamp socket shall be so fixed that the lock pins of the bulbs when installed shall lie in a vertical plane.

Any socket adjusting device requiring the dismantling of the head-lamp reflector, or the adjustment of which requires the handling of the bulb, shall be disapproved.

Any focusing device must be constructed so that neither in adjustment nor in use can the bulb base be moved out of its axial position in the reflector.

All socket adjustments must be so designed that they can be locked securely in the desired position.

Head Lamp Illumination (S. A. E.)

The head lamps shall be so arranged that no portion of the direct reflected beam cone of light, when measured 75 feet ahead of the head lamps, shall rise above 42 in. from the level surface of the road on which the vehicle stands, under any condition of loading; nor shall any portion of the direct reflected beam cone of light rise, at the 75-ft. distance, more than 12 in. above the center of the head lamp.

Head Lamp Mounting (S. A. E.)

It is recommended that head lamps be mounted so that their centers are not less than 3 ft. nor more than 4 ft. from the ground. This mounting will allow head lamps to more nearly conform to the Illumination Recommendation.

Head Lamp Brackets (S. A. E.)

Three standard sizes are recommended for the forked type of head-lamp bracket, the forks having center-to-center widths of 7¼, 8¼ and 9¼ in. The upper ends of the brackets are to be ½ in. diameter, with ½ in. S. A. E. threads and machined shoulders not less than ¾ in. diameter. The distance from the upper face of the shoulder to the last full thread on the end of the bracket should be not less than 1½ in. where no tie-rod is used, or 1½ in. plus thickness of rod where a rod is used. The use of nuts and lock washers for locking lamp to the fork is recommended.

An adjustment should be provided for the bracket to allow a change of the vertical angle of the lamp without bending any part of the bracket. The props attached to the lamp housing shall have bores of 17/32 in., the bores being 1½ in. long. The centerline of the hole in the prop shall be not less than 9/16 in. from the nearest point of the lamp housing. The clearance between the lower part of the bracket and the lamp housing shall be not less than 9/16 in.

APPENDIX

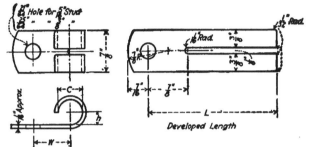

FIG. 318—STARTER CABLE LUGS (S. A. E.)

Cable Number	C	h	W	L	Max. Lengths
SB-1	17/32	17/64	15/16	2 3/16	5 ft.
SB-0	19/32	19/64	27/32	2 5/16	7 ft.
SB-00	21/32	21/64	7/8	2 1/2	9 ft.

NOTE:—Under Cable Numbers "SB" indicates Single Braid.

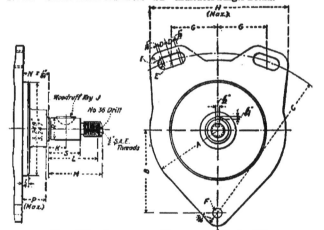

FIG. 319—GENERATOR MOUNTING (S. A. E.)

Size No.	A	B	C	D	E	F	G	H	I
1	2 3/8	2 13/16	5 3/8	1/4	7/32	19/32	1 5/8	4 3/8	7/16
2	2 5/8	3 1/16	5 1/4	5/16	1/4	19/32	1 11/16	5 3/8	1/2

Size No.	J	K	L	M	N	P	R	S
1	6	9/16	1 17/32	1 11/16	15/16	3/4	0.6250 / 0.6245	1
2	8	11/16	1 15/32	1 15/16	1 1/16	1	0.7500 / 0.7495	1 1/4

Copper Wire, Brown & Sharpe Gauge

No.	Diameter in Mils	Cross Section in Circular Mils	Resistance at 68 deg. F.	Weight in Lbs. per 1000 Ft.	Feet per Lb.
0000	460	211,600	0.0490	640	1.56
000	410	167,800	0.0618	508	1.97
00	365	133,100	0.0779	403	2.48
0	325	105,500	0.0983	319	3.13
1	289	83,690	0.124	253	3.95
2	258	66,370	0.156	201	4.98
3	229	52,640	0.197	159	6.28
4	204	41,740	0.248	126	7.91
5	182	33,100	0.313	100	9.98
6	162	26,250	0.395	79.5	12.6
7	144	20,820	0.498	63.0	15.9
8	128	16,510	0.628	50.0	20.0
10	102	10,380	0.999	31.4	31.8
12	80.8	6,530	1.59	19.8	50.6
14	64.1	4,107	2.52	12.4	80.4
15	57.1	3,257	3.18	9.86	101
16	50.8	2,583	4.01	7.82	128
17	45.3	2,048	5.06	6.20	161
18	40.3	1,624	6.38	4.92	203
19	35.9	1,288	8.05	3.90	256
20	31.9	1,022	10.1	3.09	323
21	28.5	810	12.8	2.45	408
22	25.3	642	16.1	1.94	514
23	22.6	509	20.4	1.54	648
24	20.1	404	25.7	1.22	818
25	17.9	320	32.4	0.970	1,030
26	15.9	254	40.8	0.769	1,300
27	14.2	201	51.5	0.610	1,640
28	12.6	160	64.9	0.484	2,070
29	11.3	127	81.8	0.384	2,610
30	10.0	100	103.0	0.304	3,290
31	8.93	79.7	130	0.241	4,140
32	7.95	63.2	164	0.191	5,230
33	7.08	50.1	207	0.152	6,590
34	6.30	39.7	261	0.120	8,310
35	5.61	31.5	329	0.0954	10,500
36	5.00	25.0	415	0.0757	13,200
38	3.96	15.7	660	0.0476	21,000
40	3.14	9.89	1,050	0.0299	33,400

The thickness of different insulating coverings for magnet wires is as follows: Single cotton covered, Nos. 10 to 19, 2.5-5.0 mils; Nos. 20-40, 2.0-4.0 mils; double cotton covered, Nos. 10 to 19, 5.0-10.0 mils; Nos. 20-40, 4.0-8.0 mils; single silk covered, Nos. 16-40, 1.0-2.0 mils; double silk covered, Nos. 16-40, 2.0-4.0 mils; enamel covered, Nos. 13-15, 1.5 mils; Nos. 16-28, 1.1 mils; Nos. 29-33, 0.7-1.0 mils; Nos. 34-36, 0.4-0.7 mils; Nos. 38-40, 0.3-0.6 mils.

INDEX

Alternating mains, charging from, 213
Ammeter, principle of, 62
Armature bands, 71, 244
Armature construction, 240
Armature core, magneto, 71
Armature reaction, 73, 253
Armature winding, 243
Armature winding diagrams, 245
Atwater Kent interrupter, 121, 128
Atwater Kent spark governor, 174
Automatic change gear, Jesco, 341
Automatic ignition switch, 123
Automatic timing, 171
Automatic timing coupling, 173
Auxiliary spark gap, 103
Back-kicks, provision against, 176
Bakelite, 162
Bar magnet, field of force of, 42
Battery and coil system, 84
Battery capacity required, 377
Battery charging, economies of, 213
Battery charging installations, 213
Battery cut-out, centrifugal, 288
Battery cut-out, Delco, 291
Battery cut-out, Entz, 286
Battery cut-out, electro-magnetic, 290
Battery cut-out, Holtzer-Cabot, 288
Battery cut-out, location of, 293
Battery cut-out, mercury type, 289
Battery cut-out, Mira, 288
Battery cut-out, need of, 259
Battery cut-out types compared, 292
Battery floating on line, 260
Battery holders, wireless, 15
Battery instructions, S. A. E., 31
Battery mounting, 378
Battery savers, 123
Battery spring-mounting, 379
Battery systems, modern, 119
Batteries, care of, 381
Batteries not holding charge, 382
Batteries, standard dimensions of, 402
Batteries, winter care of, 383
Baume scale, 26
Bendix drive, 348
Bijur automatic shift, 350
Black oxide of manganese, 14
Bosch dual system, 182
Bosch duplex system, 184
Bosch high-tension magneto, 143

Bosch magneto for V engines, 160
Bosch "Two Independent" systems, 177
Brushes and rigging, 251
Brushes, care and adjustment of, 255
Brushes, sanding in, 257
Buckproof coils, 109
Bulb bases, 306
Bulb sizes, S. A. E., 395
Bulbs, candle power of, 303
Bulbs, life and efficiency of, 303
Bulbs, nitrogen-filled, 302
Cable connectors, 200
Cable specifications, S. A. E., 397
Cable terminals, 200
Cable terminal, single wire, 308
Cable testing device, 227
Candle power, 296
Candle power, mean horizontal, 305
Candle power, mean spherical, 304
Canfield spark plug patent, 101
Capacity, 57
Charging indicator, U. S. L., 293
Charging indicator, Gray & Davis, 294
Charge determination, 25
Charging storage batteries, 25
Charging voltages, 211
Chemical effects of current, 4
Closed circuit interrupters, 119
Clutches, overrunning, 343
Coils, care of, 117
Coils, locating faults in, 117
Coils, construction of, 87
Combed pole tips, 169
Combined magneto and battery systems, 177
Commutation, 253
Commutator, 239
Compound magnets, 45
Compound-wound generator, 249
Compound-wound motor, 325
Condenser, 57, 89
Conductors, 2
Connecticut interrupter, 122
Connection of instruments, 66
Connector plugs, 307
Contact points, dressing up, 116
Contact points, life of, 122
Counter electromotive force, 326
Core disks, 241
Coupling boxes, 374

INDEX

Cranking speed, 329
Current lag, 55
Current control, 272
Current consumption of ignition systems, 130
D'Aessouval type instrument, 62
Dashboard instruments, 65
Dash lamps, 316
Delco interrupter, 125, 128
Delco distributer, 111
Delco relay, 125
Depolarizers, 13
Differential field windings, 250, 273
Dimming devices, 318
Direct current mains, charging from, 211
Discharge rate, effect on capacity, 24
Discharge rate, effect on voltage, 25
Distributer, 109
Distributer mounting, 395
Direction of current, rule for, 10
Direction of e.m.f., rule for, 54
Dixie magneto, 157
Dixie magneto, spark advance, 170
Dry cells, 13
Dry cells, deterioration of, 14
Dry cells, reviving old, 16
Dynamo generator, principles of, 236
Dynamos, care of, 255
Dynamotors, double deck, 342
Energy consumed in starting, 329
Edison battery, care of, 38
Edison battery, charging characteristics of, 37
Edison battery, voltage of, 37
Edison storage battery, 34
Eisemann automatic timing mechanism, 171
Eisemann helical pole faces, 168
Electric energy, 8
Electricity, current, 1
Electricity, nature of, 1
Electricity, static, 1
Electrolyte, adjusting level of, 28
Electrolyte, mixing, 28
Electrolyte of Edison battery, 37
Electrolyte, resistivity of, 30
Electrolytes, 4
Electro-magnetic induction, 49
Electrostatic interference, 207
Esterline headlamp, 310
Exide storage cell, 22
Extra current, absorption of, by condenser, 59
Faraday's laws of induction, 50
Faure storage cells, 17
Firing order, 108
Field circuit, open, 389
Field connections, 247
Field frames, 237
Filament development of metal, 299
Filaments, types of, 301
Flexible steel tubing, S. A. E., 400
Focusing devices, types of, 312
Focusing headlamps, 310
Ford flywheel magneto, 132
Ford ignition system, connections of, 135
Ford magneto, magnetic circuit of, 132
Ford system, characteristics of, 134
Form-wound armature, 244
Freezing temperatures of electrolyte, 384

Fuse blocks, 361
Fuses, 360
Galalith, 162
Gaskets, spark plug, 96
Generator capacity required, 254
Generator characteristics, 250, 260
Generator control, 259
Generator control, Adlake, 267
Generator control, Bijur, 270
Generator control, Bosch, 265
Generator control by armature reaction, 276
Generator control, Delco, 267
Generator control, magnetic shunt, 276
Generator control, magnetic vibrator, 269
Generator control, Rushmore, 273
Generator control, third brush, 277
Generator control, U. S. L., 265
Generator control, thermostatic, 283
Generator control, revise series, 272
Generator control, Westinghouse, 279
Generator, electrical defects in, 387
Generator mounting, S. A. E., 405
Generator output, effect of temperature, 282
Generators, 235, 260
Glare, 316
Governor clutches, 263
Gray & Davis governor clutch, 263
Gray & Davis headlamp, 309
Grids, 18
Ground return system, 365
Grounds, 66
Hand starting magneto, 187
Hand-wound armature, 244
Hard rubber, insulating properties of, 161
Headlamp brackets, 404
Headlamp focusing, 404
Headlamp mounting, 313, 404
Headlamp nomenclature, 393
Headlamp sockets, S. A. E., 404
Headlight tests, specifications for, 391
Helical slots, spark timing by, 171
Helix or coil, action of, 39
Herz governor coupling, 173
Heat losses in conductors, 8
High-tension ignition, 83
High-tension magneto, current wave of, 151
High-tension magneto, flux curve of, 151
High-tension magneto, invention of, 143
High-tension magneto, principle of operation of, 150
High-tension magneto, voltage curve of, 151
History of electric starters, 235
Hitenagraph, 230
Horseshoe magnet, 45
Humming, prevention of, 254
Hydraulic analogy, 2
Hydrometer tests, 382
Hydrometer syringe, 27
Hysteresis, 47
Illumination and glare tests, 322
Illumination, control of, 318
Illumination, law of, 296
Ignition cable, 200
Ignition locks, 205

INDEX

Ignition, low tension, 76
Ignition timing, 113
Impulse starters, 188
Individual coils, 107
Induction coil, 83
Induction coil, resistance of, 88
Induced currents, direction of, 51
Induction between adjacent coils, 49
Induction, law of, 50
Inspection lamps, 316
Installation specifications, S. A. E., 371
Instrument shunt, 64
Instruments, use of, 68
Insulated return system, 365
Insulation requirements, S. A. E., 403
Insulators, 2
Insulators, high tension, 161
Insulators, spark plug, 96
Intensifiers, 319
Interference, electrostatic, 207
Interrupter, adjustment of magneto, 164
Interrupter, Bosch magneto, 147
Interrupters, high speed, 127
Iron wire resistance characteristics, 275
Junction boxes, 374
K-W magneto, 156
K-W impulse starter, 190
Lag of interrupter, 123
Lamp voltage, 302
Lamp sockets, 307
Lamps used, types of, 305
Lens, Osgood, 321
Lens, McKee, 321
Lenses, prismatic, 321
Light, unit of, 296
Lines of force, 39
Low tension ignition, 76
Magnets, aging, 47
Magnet frame, 70
Magnets, lifting power of, 47
Magnet material, 46
Magnetic circuit, the, 41
Magnetic circuit, law of the, 42
Magnetic meshing gear, 354
Magnetic plug system, 81
Magnetic spark plugs, 80
Magnetism, nature of, 39
Magnetizing steel, 46
Magneto and coil ignition, 132
Magneto, care and adjustment of, 163
Magneto dimensions, S. A. E., 402
Magnetos for two-point ignition, 193
Magnetos for V-type engines, 159
Magneto generator, construction of, 70
Magneto generator, principle of, 71
Magnet recharging, 222
Magneto tester, Butler, 228
Magneto, distributerless, 197
Magneto, size of, 74
Magneto troubles, location of, 164
Magneto starting helps, 186
Magneto, standard dimensions, 75
Magneto voltage, 72
Make-and-break induced E. M. Fs., 84
Make-and-break mechanism, 77
Master vibrator, 112
Mea magneto, 169
Metric spark plug, 99

Mica, under-cutting, 257
Motor characteristics, 324
Motor generator, 221
Multi-cylinder coil ignition, 107
Multi-vibrator coil systems, 108
Mutual induction, 60
Nitrogen-filled bulbs, 302
Non-inductive coil, 53
Non-sooting recess plugs, 101
Non-vibrator coils, 130
Ohm's law, 3
Open circuit interrupters, 119
Oscillograph diagrams, 94
Oscillographs, 231
Oscillogram of starter current, 331
Parabolic reflector, 297
Parallel connection, 12
Peaked current curve, reasons for, 167
Permanent magnets, 44
Permanent magnets, types of, 40
Permeability, 43
Philbrin interrupter, 129
Photometer, 297
Pittsfield aircraft spark plug, 105
Planté storage cells, 17
Polarization, 11
Porcelains, protected, 100
Power required in starting, 326
Pocket instruments, 65
Polarity of mains, determining, 212
Potential, difference of, 2
Primary batteries, nature of, 10
Rectifier, electrolytic, 214
Rectifier, mercury arc, 218
Rectifier, Tungar, 219
Rectifier, vibrator type, 215
Reflection, law of, 297
Reflectors, 297
Reflectors, care of metal, 313
Reflector displaced, halves of, 320
Reflectors, glass, 309
Reflectors, metallic, 308
Regulators, care of, 284
Remy dual system, connections of, 140
Remy inductor magneto, 154
Remy Model P magneto, 137
Remy thermostatic control, 283
Resistance calculations, 6
Resistance, specific, 5
Resistance, starter, 347
Reverse series control, 272
Rocking magnet frame, 169
Rotary sector type magneto, 153
Safety spark gap, 145
Self-excitation, 246
Self-induction, 51
Semi-automatic impulse starters, 190
Series connection, 12
Series generator, 248
Series motor, 324
Series multiple connection, 12
Shifting gear drives, 345
Short circuits, 66
Shunt generator, 249
Shunt motor, 324
Sine wave, 53
Single coil system, 109
Single unit systems, 334
Single vs. multiple sparks, 119
Spark coil, 76
Spark plug, elements of the, 96
Spark plug faults, 102
Spark plugs for aircraft engines, 105